Chap. II
85-86

The Basic Book of
Electricity and Electronics

Gerald E. Williams

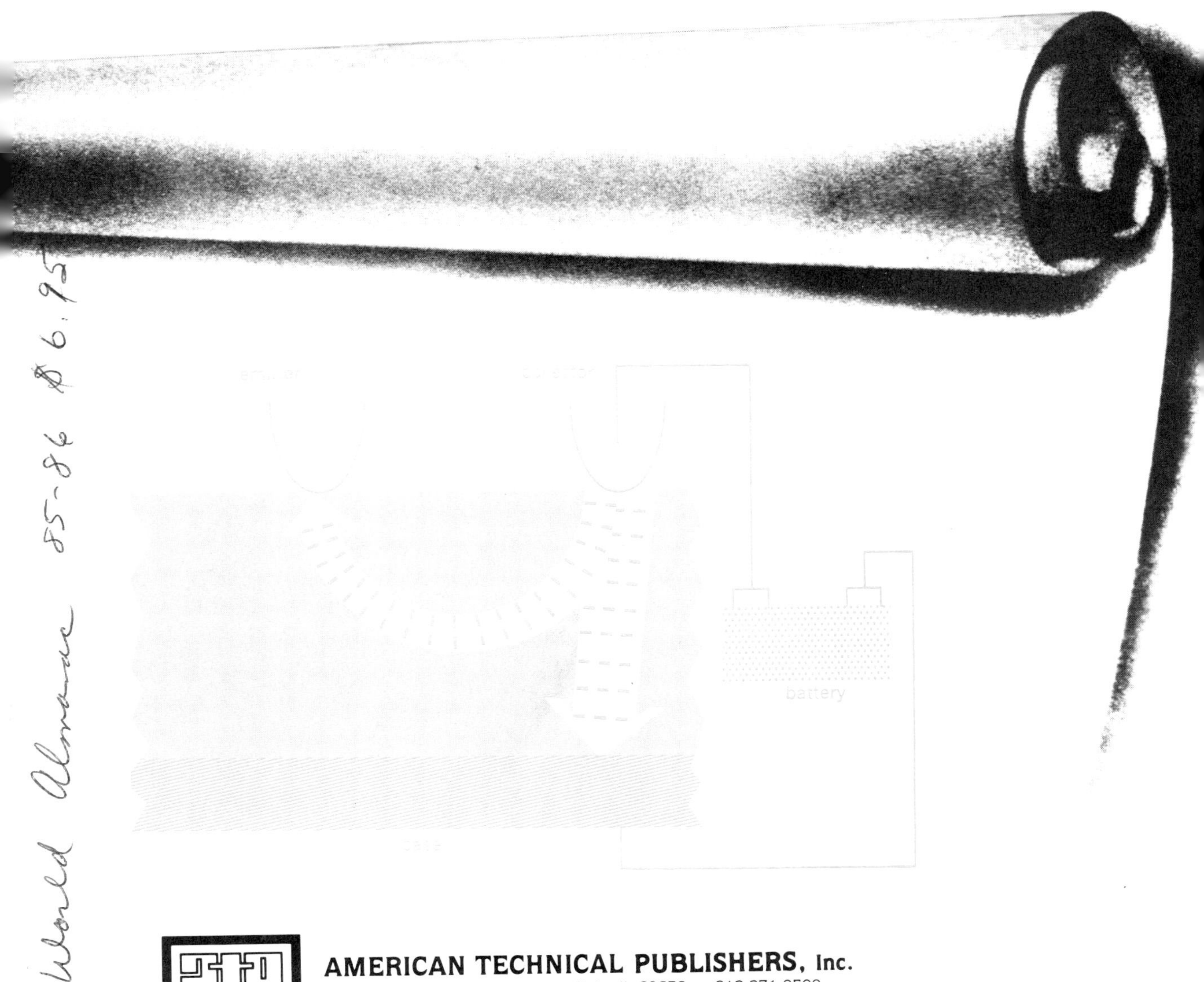

AMERICAN TECHNICAL PUBLISHERS, Inc.
12235 S. Laramie Ave. • Alsip, IL 60658 • 312-371-9500

Library of Congress Catalog Number: 79-51401
ISBN: 0-8269-1485-3

123456789-79-98765432

To my mother-in-law and friend
Lenore Gibson

PRINTED IN THE UNITED STATES OF AMERICA

CONTENTS

PREFACE

THE BASIC BOOK OF ELECTRICITY AND ELECTRONICS is part of an integrated series of Industrial Arts textbooks designed to teach basic skills to beginning students. Its major objectives are: career exploration, development of consumer awareness, manipulative skills, and craftsmanship. The philosophy of THE BASIC BOOK OF ELECTRICITY AND ELECTRONICS is based on a recent, nationwide survey in which electricity and electronics teachers at all levels were asked to outline the courses they taught and let us know what types of instructional materials they actually needed. The result is a highly-visual text with a controlled reading level that will help insure student success.

The author and the publisher wish to acknowledge and thank the following individuals for their assistance and cooperation: Patty Williams, Kelly Williams, Geoffrey Williams, and Judy Anderson.

The Publisher

INTRODUCTION 1

You are about to take a trip into the exciting world of electricity and electronics. You will learn how to make printed circuit boards and how to build projects straight out of space age technology. You will learn about the different electronic parts and how they connect together to make radios, stereos, and computers work.

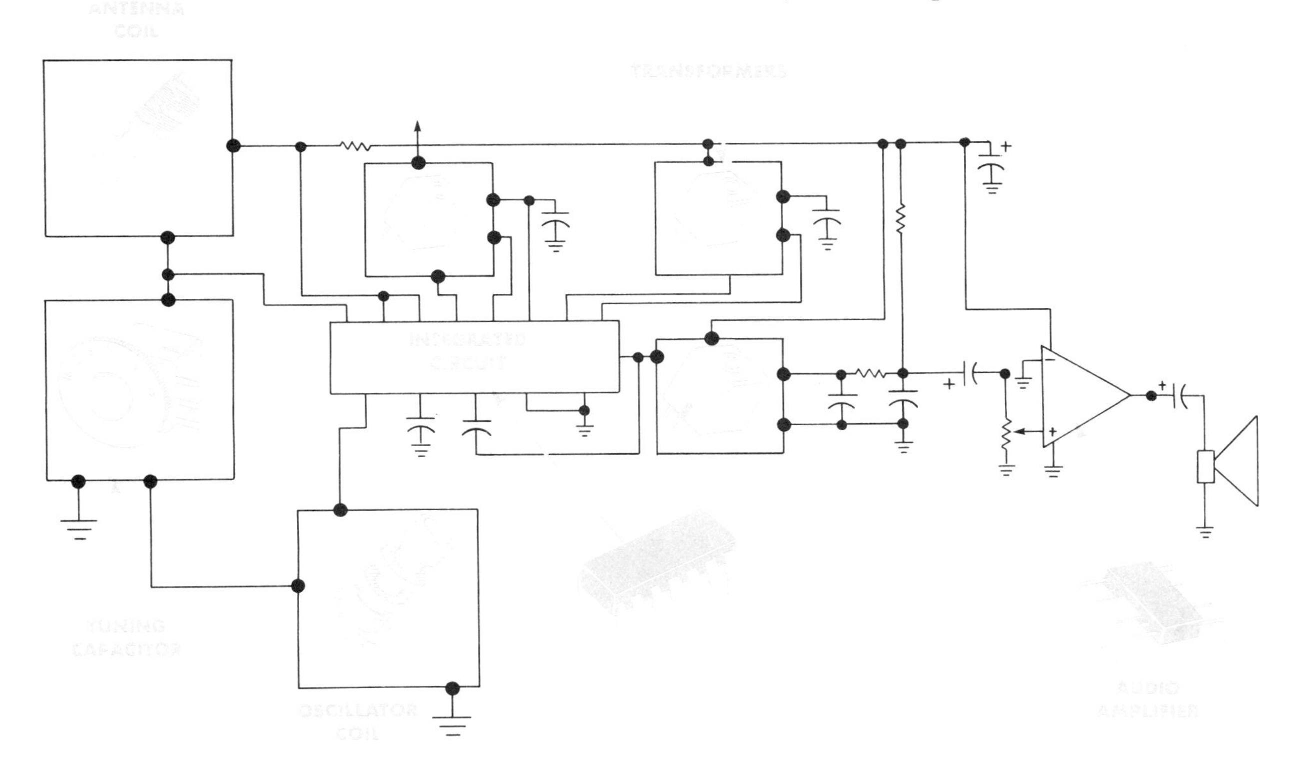

Electricity is everywhere. Electric lights, motors, color televisions, microwave ovens, and stereo systems are as common as tables and chairs.

When the transistor was invented at Bell labs in the 1940's a revolution in electronics was started.

The space program's need for miniature electronics helped bring the integrated circuit into the world. Thousands of transistors and other parts can be put on a chip of silicon less than 1/4″ square. Complete computers can be put in packages no bigger than a large postage stamp. The cost of a computer has dropped from hundreds of thousands of dollars to $50 or less.

You can find work in electricity or in electronics almost anywhere. A commercial jet is a flying electronics laboratory. It has computers, radios, navigational equipment, electric motors, even stereo systems.

There is also a big need for electronics technicians in the automotive service industry. The electrical system in automobiles is getting very complicated. On-board computers and computer technicians will be a part of auto technology from now on.

Installation and repair of home electronics equipment is a major industry—and growing. Even surveying crews have traded in their rods and telescopes. They now use laser instruments with built-in computers to survey.

Hospital electronics equipment diagnoses and monitors the condition of

IN YOUR CAR

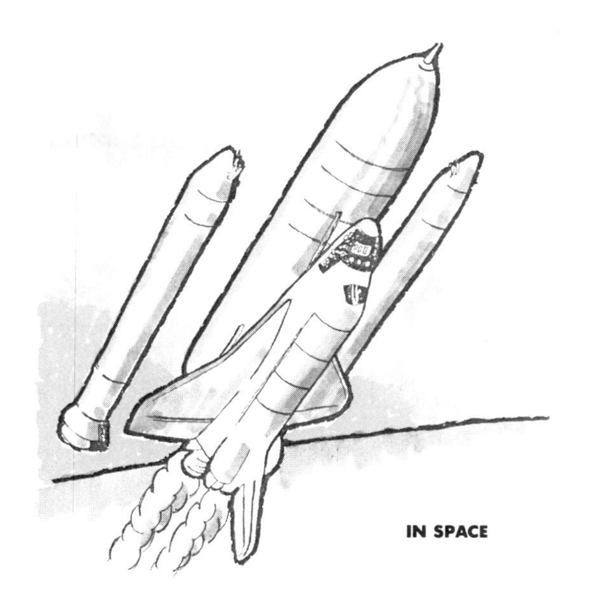

IN SPACE

patients. Pacemakers keep people alive who have certain heart problems.

Engineers and technicians are needed to design and develop new electronics systems. New systems, like those mentioned, are being designed and worked on everyday. The end is nowhere in sight.

The list of electronics opportunities is very long and growing. Careers in electronics are almost unlimited. Each year electronics finds its way into places where it has never been used before.

IN CAMERAS

ON THE STREET

If you are thinking about a career in electronics you may need to know quite a bit of math. Some people find math easy. Others struggle with it for a long time before they can make any sense out of it. But struggle long enough and math will make sense to you.

SELF CHECK

1. What happened at Bell labs in the 1940's?
2. What has the space program added to the field of electronics?
3. What electronics equipment is used on an automobile?
4. What basic skill will you need for a career in electronics?

IN YOUR HOME

Unit 2 SAFETY

An accident can take all the fun out of your project building. Here are some safety rules to help keep you out of trouble. Read these rules, add a few pounds of common sense, and you will stay healthy and happy.

- ALWAYS WEAR EYE PROTECTION WHEN YOU ARE OPERATING ANY KIND OF POWER TOOL.
- CLAMP ALL WORK TO BE DRILLED SO IT CANNOT TURN IF THE DRILL STICKS.
- DON'T LEAVE THE CHUCK KEY IN THE CHUCK OF A DRILL PRESS OR DRILL MOTOR.
- WHEN USING A SCREWDRIVER OR OTHER SHARP TOOL, BE SURE TO KEEP YOUR HAND WHERE IT WON'T BE STABBED IF THE TOOL SLIPS.
- THE CUT EDGE OF SHEET METAL IS RAZOR SHARP — BE CAREFUL.
- KEEP YOUR WORK AREA CLEAN.

Figure 2-1: Use common sense, and you'll be a safe worker.

- NEVER WORK ON ELECTRICAL EQUIPMENT WHEN YOU ARE ALONE.
- NEVER WORK ON ELECTRICAL EQUIPMENT WHEN YOU ARE TIRED.
- TRY NOT TO WORK ON ELECTRICAL EQUIPMENT WHILE IT IS PLUGGED-IN.
- ALWAYS KEEP ONE HAND BEHIND YOUR BACK WHEN WORKING ON LIVE ELECTRICAL EQUIPMENT EVEN LOW VOLTAGE EQUIPMENT.
- DO NOT TOUCH ELECTRICAL EQUIPMENT IF YOU ARE STANDING ON A DAMP FLOOR — OR A METAL FLOOR.
- DO NOT TOUCH ELECTRICAL EQUIPMENT IF YOU OR THE EQUIPMENT IS DAMP OR WET.
- NEVER USE WATER TO PUT OUT AN ELECTRICAL FIRE.
- KNOW WHERE THE FIRE EXTINGUISHER IS — AND HOW TO USE IT.
- DO NOT TOUCH TWO PIECES OF PLUGGED-IN EQUIPMENT AT THE SAME TIME.
- DO NOT TOUCH A WIRE FROM THE POWER LINE UNLESS YOU HAVE CHECKED TO MAKE SURE IT IS "DEAD".
- CHECK ALL "DEAD" CIRCUITS BEFORE YOU TOUCH THEM.
- DO NOT HOLD SOLDER IN YOUR MOUTH. THE SOLDERING IRON TIP MAY SHORT YOU TO THE POWER LINE.
- NEVER TAKE A SHOCK ON PURPOSE.
- TURN OFF ALL POWER, AND GROUND ALL HIGH VOLTAGE POINTS BEFORE YOU WORK ON HIGH VOLTAGE ELECTRONIC CIRCUITS. A CAPACITOR CAN STORE A FATAL CHARGE.

Figure 2-2: Electricity is dangerous as well as helpful. Follow these simple rules, and you may live a long and happy life.

SELF CHECK

1. When should you wear eye protection?
2. Why is it dangerous to hold solder in your mouth?
3. Why should you cup your hand around a wire before cutting it?
4. List three safety rules that should be observed while working with "live" equipment.

Learning something new usually means learning about some special tools. In electronics and light electrical work the basic tools are:

- Screwdrivers
- Crescent wrenches
- Pliers
- Hex nut drivers
- Wire strippers
- Crimping tool
- Soldering iron

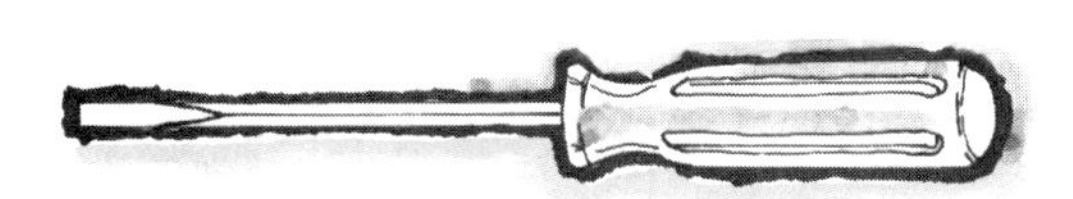

FLAT BLADE SCREWDRIVER

Figure 3-1: Flat blade screwdrivers are measured across the blade tip. The flat blade sizes most often needed for electrical work are: 1/8″ and 3/16″.

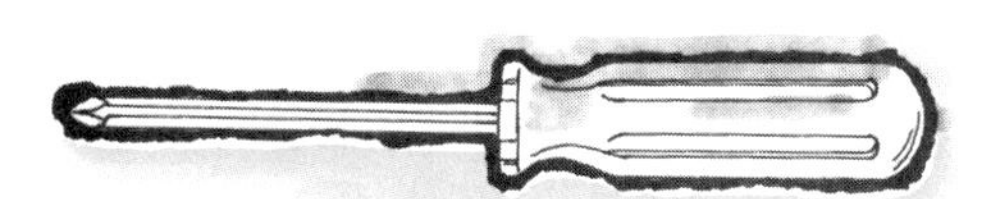

PHILLIPS HEAD SCREWDRIVER

Figure 3-2: Phillips (cross point) screwdrivers have a number for each size. The commonest sizes are: #0, #1, and #2.

CRESCENT WRENCH

Figure 3-3: A 5 or 6 inch long adjustable wrench does the job of several ordinary wrenches. An adjustable wrench will also fit metric nuts.

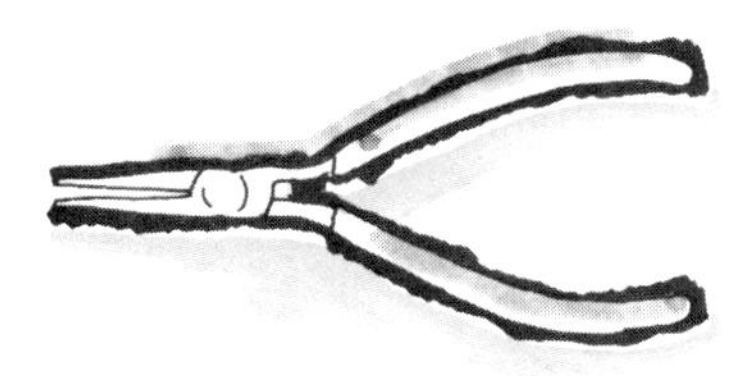

LONG NOSE PLIERS

Figure 3-4: You will also need a pair of long-nose pliers. Four and a half to 5 inches is a good length.

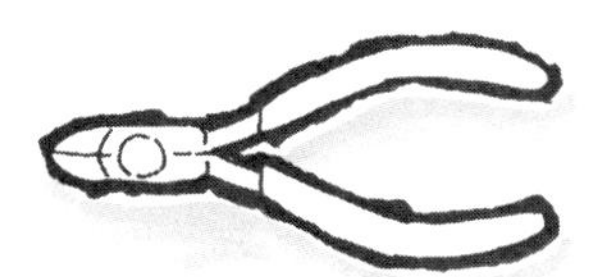

DIAGONAL CUTTING PLIERS

Figure 3-5: Diagonal cutters, called dikes, are used to cut wire. The most useful length is 4-1/2 to 5 inches.

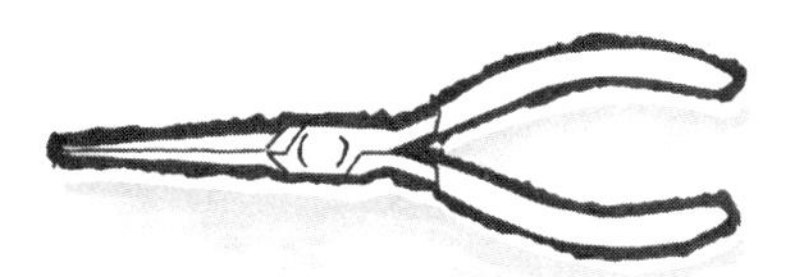

NEEDLE NOSE PLIERS

Figure 3-6: Needle-nose pliers are like long-nose pliers except that the nose is longer and thinner.

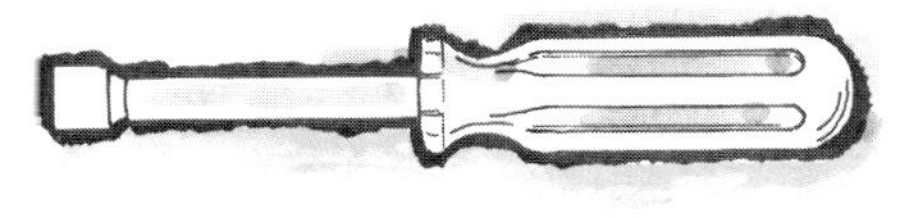

HEX NUT DRIVER

Figure 3-7: A hex nut driver set is handy. But you can get along without it.

SOLDERING PENCIL (IRON)

Figure 3-8: The soldering pencil (iron) is one of the most important of all tools in electronics. In modern electronics work, a 35 to 40 watt iron is about the biggest one that should be used.

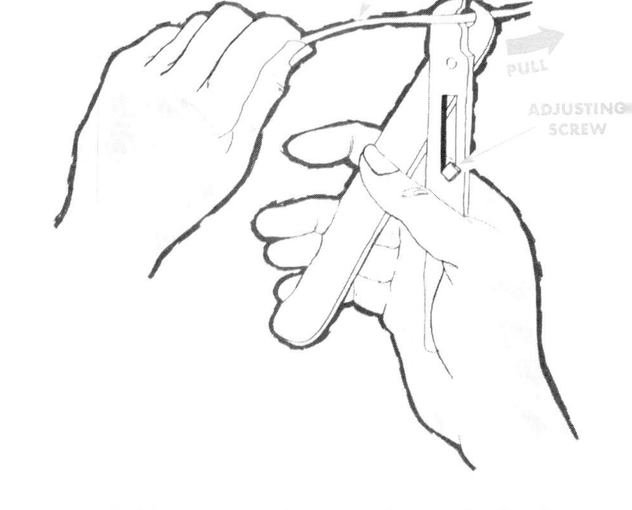

Figure 3-10: A wire stripper makes stripping insulation easy and prevents nicks in the wire. Nicked wires break very easily. Notice the screw in the slot in the right handle of the stripper. This screw must be adjusted for the size wire you are using.

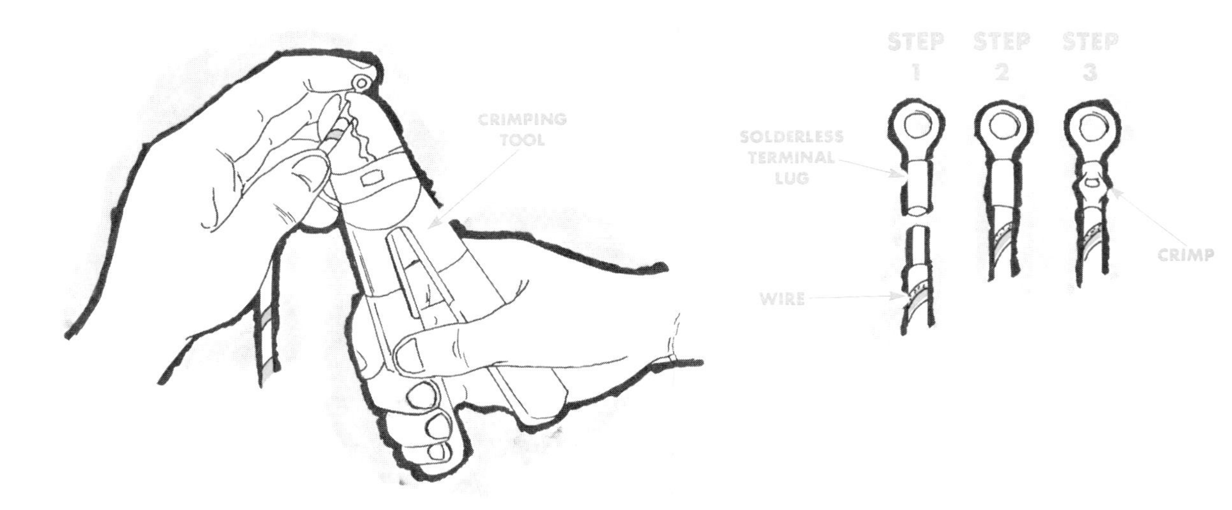

Figure 3-9: The crimping tool is used to attach solderless terminals to wire.

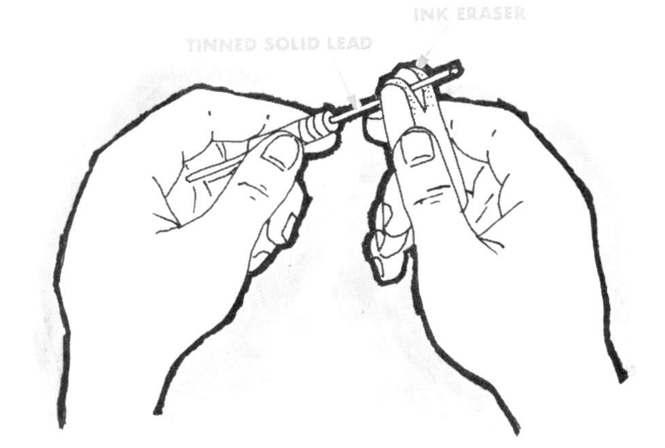

Figure 3-11: You can make a useful tool out of an ordinary ink eraser. It's used for cleaning the wires on electronics and electrical parts. These wires tarnish and become hard to solder. To make the tool, cut a V-shaped slot in one end of an ink eraser. The uncut end can be used to clean the copper on printed circuit boards.

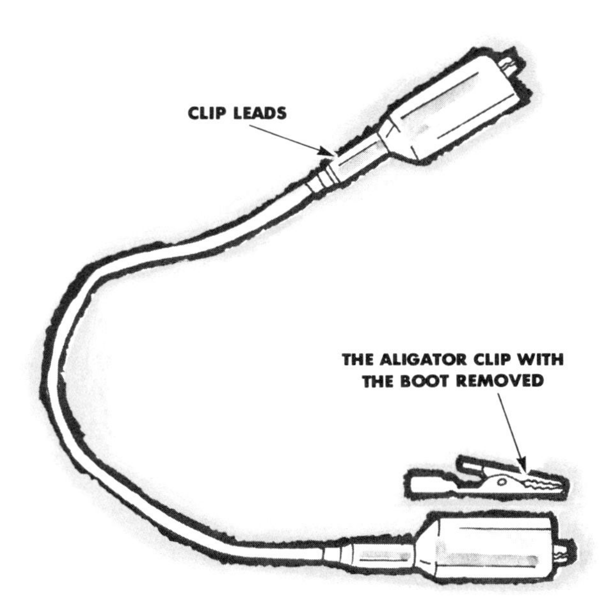

Figure 3-12: A final item that should be in your tool kit is a dozen clip leads. Twelve to 18 inches is a good length. The clips on the ends of the wire should be covered by plastic boots to prevent shorting.

SELF CHECK

1. List two types of screwdrivers.
2. Why is an adjustable wrench useful?
3. What does a nick do to a wire?
4. Why are plastic boots used on clip leads?

2 DOING IS LEARNING

Soldering takes practice and a good way to practice is to build something. If your solder joints are good the project will work.

The coin toss game on this page, will provide you with soldering practice and fun afterward.

The object of The Coin Toss game is to land coins onto targets. A coin on target lights the light and sounds the buzzer. The drawings show how to hook up the circuit. Be sure you make good solder joints.

Test the circuit by placing a coin on any target.

A LED (light emitting diode) must be connected the right way. If it doesn't light, unsolder it and turn it around.

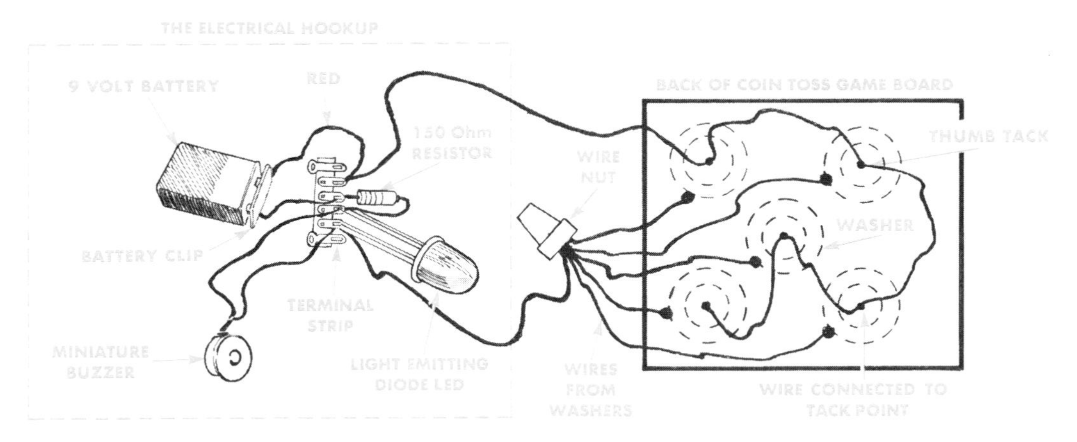

COIN TOSS GAME BOARD: The tossed coin completes the circuit between the thumbtack and the washer. The washers are glued in place on the circuit board, and a hole is drilled through the board next to them so the wire can be passed through and soldered to the washer.

WIRING DIAGRAM: Wires are soldered to the thumbtack points on the back of the board. The wires that attach to the washers are then passed through the holes and soldered to the washer on top of the board.

Soldering is simply the joining of two or more wires with melted solder (figure 4-1).

To get good solder joints:

- Start with the right mechanical joint (figure 4-2)
- Use the right size soldering tool (figures 4-3 and 4-4)
- Use the right solder

Acid core solder is never used in any kind of electrical work! What the acid doesn't eat away it will short out. Special multicored solders meant for electronics soldering are the best for all electrical work.

Some wire is made up of many strands of fine wire. Stranded wires should be twisted together so none of the strands stick out. Then flow just enough solder to hold the strands together.

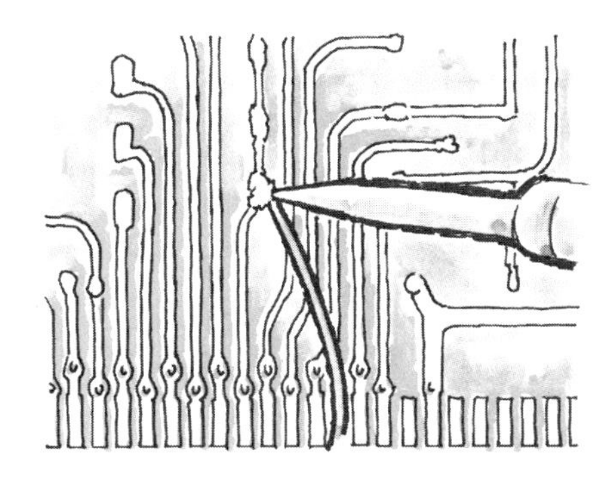

Figure 4-1: The best way to make sure an electronic project won't work is to make poor solder joints. The only way to be sure of a good solder joint is to do it right the first time. Most of the time you can't look at a solder joint and tell whether it is good or bad. In this unit we will learn the secrets of good soldering.

SAFETY NOTE: Always wear safety glasses while soldering. Many wires used in electronics are made of springy material that can flip molten solder into your eyes.

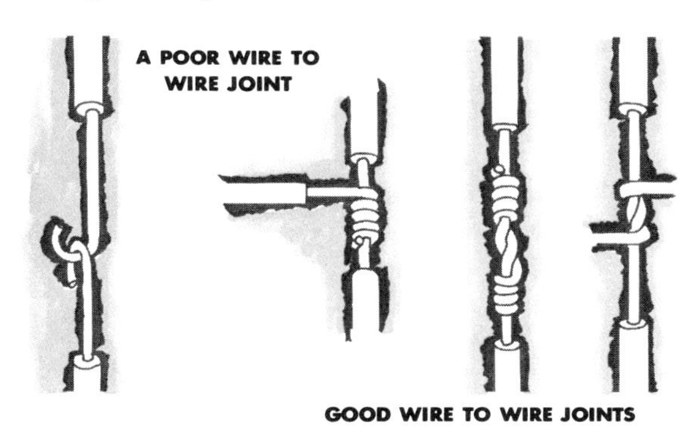

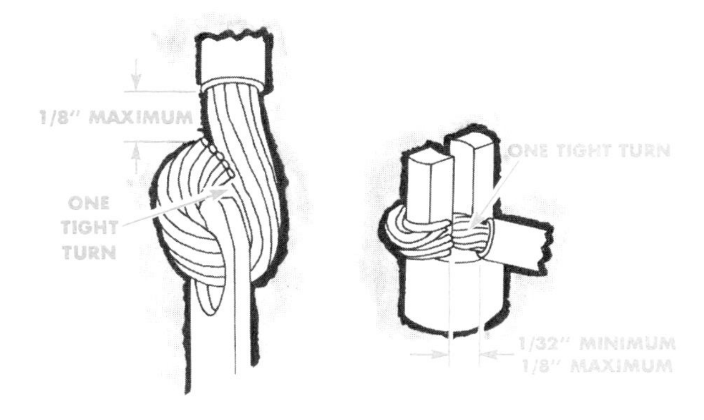

Figure 4-2: The start of any good soldering joint is a good mechanical joint.

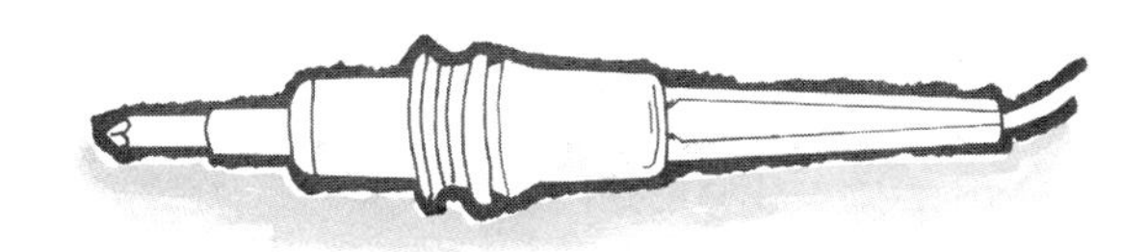

Figure 4-3: Soldering pencils rated at 25 to 35 watts are just right for most electronics work.

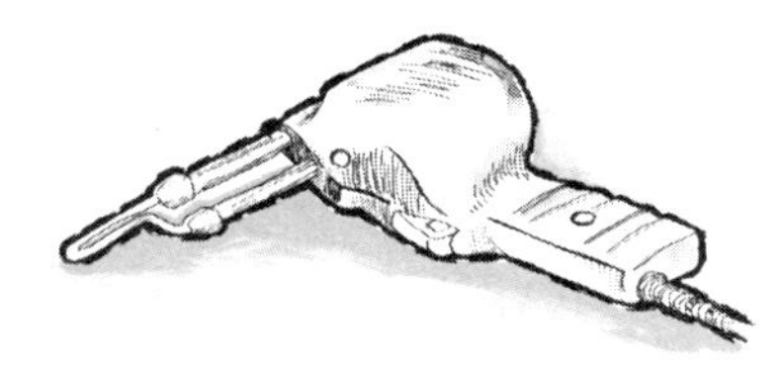

Figure 4-4: Soldering guns, rated at about 120 watts, are great for household electrical repair work.

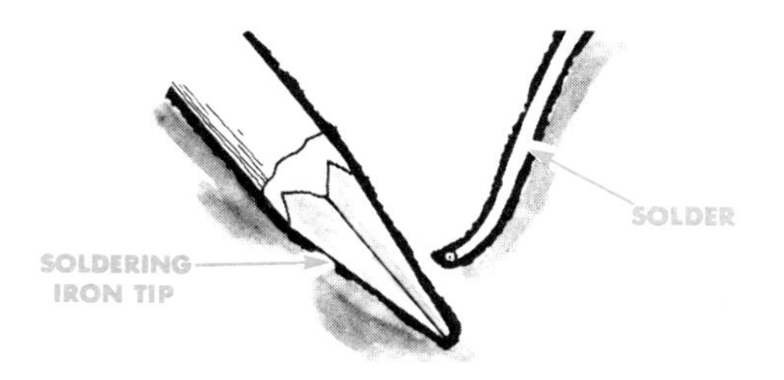

Figure 4-5: Soldering is easy if you keep the soldering iron tip clean with a thin coating of clean solder on the tip. This is called tinning.

Figure 4-6: A damp sponge does a good job of tip cleaning.

SELF CHECK

1. What size soldering pencil is good for most electronics work?
2. What size soldering gun is good for household electrical work?
3. What kind of solder is best for electrical and electronics work?
4. Why is it important to wear safety glasses?

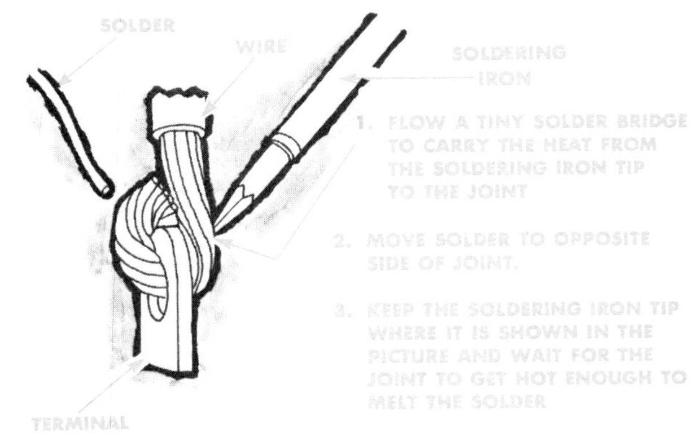

Figure 4-7: Soldering is easy if you follow a few simple rules.

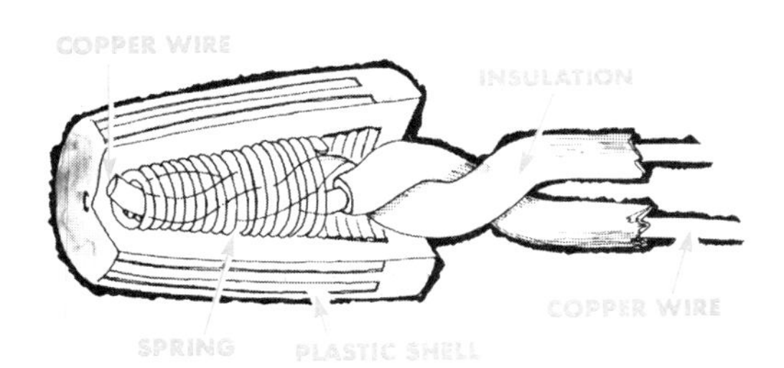

Figure 4-8: The wire nut is meant to be used without soldering. Where the joint will be permanent, soldering the joint before putting on the nut is a good idea. Just don't use too much solder or the wire nut won't stay on.

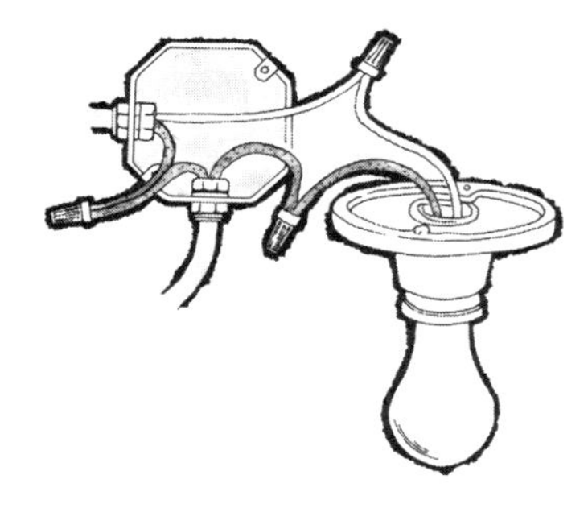

Figure 4-9: Wire nuts are often used in wiring household electrical circuits. They are often handy in electronics projects too.

MATERIALS FOR A PRINTED CIRCUIT BOARD — Unit 5

An important part of any electronic circuit is the circuit board.

The circuit board holds the parts in place to prevent broken wires or shorts. Many electronic parts are also easily broken if they are moved around after they are connected.

Instead of wires, the printed circuit board uses flat strips of copper glued to a plastic sheet. The blank circuit board

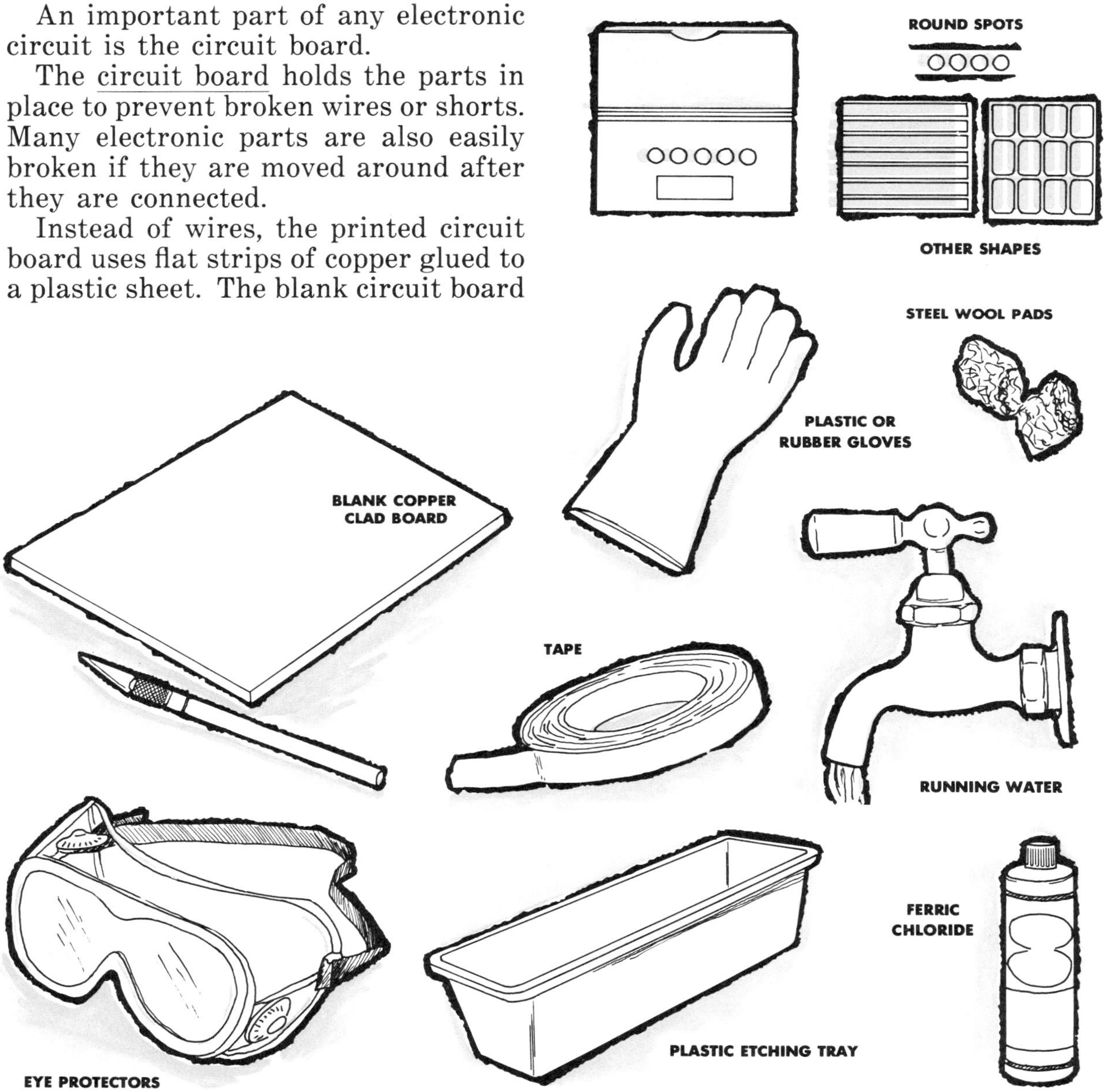

ROUND SPOTS
OTHER SHAPES
STEEL WOOL PADS
PLASTIC OR RUBBER GLOVES
BLANK COPPER CLAD BOARD
TAPE
RUNNING WATER
FERRIC CHLORIDE
PLASTIC ETCHING TRAY
EYE PROTECTORS

Figure 5-1: You will need these supplies and equipment to make a printed circuit board.

starts as a sheet of copper covering the surface of a piece of plastic board.

Lay out the circuit by placing tape on the blank circuit board where you want the wires to run, and stick adhesive spots at the ends of the tape. These spots are called "pads." The electronic parts will be soldered to these pads.

Next, the taped board is put into a chemical that eats away all the copper that is not covered by spots and tape. What is left are copper strips and pads on a plastic board.

The materials needed to make a printed circuit board are shown in figure 5-1. They are:

- Copper clad blank board (available at any electronics parts supply store or through a catalog)
- Tape and spots (Use 1/8 inch wide self-adhesive correction tape and 1/4 inch self-adhesive spots from any stationery store. Don't use labels or tape that are removable. One-eighth inch wide pin striping tape used to decorate cars is excellent. Any waterproof tape will work. Note that labels from the stationery store may float off in the chemical. But the glue will stay and keep the copper covered.)
- Etching solution (a chemical called ferric chloride)
- Etching tray (should be plastic or glass, a little bigger than the board)
- Steel wool (used for cleaning the copper)
- Pair of rubber or plastic gloves
- Goggles or face shield
- Utility knife (used for cutting the tape)

In an emergency, you can make your own copper clad board with Formica® scraps, epoxy glue and tooling copper. The tooling copper is available in hobby shops.

Ferric chloride is not an acid and will not eat holes in your clothes or skin. However, it does make a brown stain that will not wash out of clothes. To handle it safely, wear plastic or rubber gloves and goggles.

If you get it on your skin wash it off immediately with plenty of water. It will wear off in a week or two.

Ferric chloride also eats other metals, so do not pour it down the drain. It will eat the pipes. Don't use ferric chloride near tools. The fumes will rust them. It is okay to use a sink to rinse a board after you take it out of the chemical—if you use plenty of water.

Of course, you don't drink ferric chloride. If you accidentally swallow some or get it in your eyes, get medical help right away.

SELF CHECK

1. What are circuit boards used for?
2. List the materials you will need to make your own printed circuit boards.
3. What should you do if you accidently swallow the etching solution?
4. What damage will ferric chloride cause if used improperly?

Before beginning the printed circuit board, carefully clean the copper surface with steel wool. Then transfer the pattern to the blank copper clad board.

To transfer the pattern, put a sheet of onion skin or tracing paper over the layout. Make a small pencil mark at the center of each spot.

After marking the spot locations, tape the tracing paper to the copper side of the blank board. Use a center punch or sharp nail to make a punch mark where each spot should be.

Take the paper off the board. Pencil "X" marks through the center of each punch mark. Stick a self-adhesive spot at each punch mark. Use the "X" marks to center the spots.

Following the circuit board layout, connect the dots with tape.

Go over and around the tape joints with a dull pencil to make sure the tape is pressed down tight enough to keep the chemicals from getting under the tape.

Use an eraser to clean any pencil marks off the bare copper. Pencil marks on the tape are okay.

IMPORTANT: Have somebody check your layout before you etch it. Mistakes are hard to correct later.

Wearing plastic or rubber gloves, pour about 1/2 to 3/4 inch of echant solution (ferric chloride) into a plastic tray. Use a tray that is slightly larger than the board.

After 30 minutes, take the board out of the echant and rinse it carefully with water. Be sure to wear your plastic or rubber gloves.

If any copper is left outside or around the adhesive, put the board back into the echant soup. When no copper shows, wash the board with water and dishwashing detergent.

Peel off the tape and spots and clean the copper wiring with steel wool.

Drill a hole in the center of each copper circle. These circles are called pads. Use a 1/8″ drill.

The board is ready to wire.

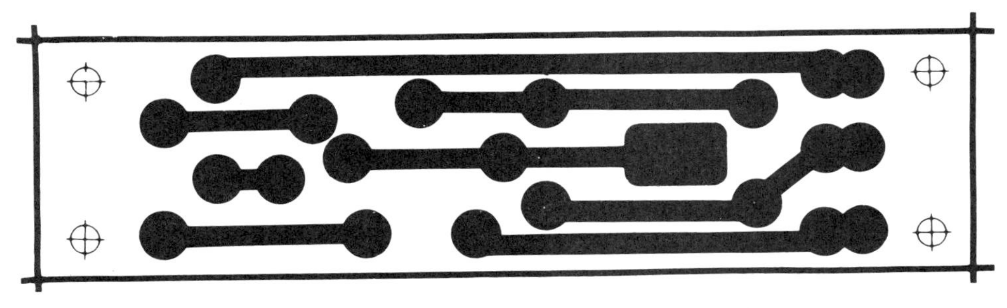

Figure 6-1: Here's a circuit board for you to make. In the next unit we will show you how to put it all together into a nifty circuit tester.

1 CLEAN THE COPPER WITH STEEL WOOL

2 CAREFULLY MARK THE CENTER OF EACH SPOT ON A SHEET OF TRACING PAPER. DON'T HURT THE BOOK.

3 TAPE THE TRACING PAPER TO THE BLANK BOARD, ON THE COPPER SIDE.

USE A SHARP CENTER PUNCH TO MARK THE LOCATION OF EACH SPOT.

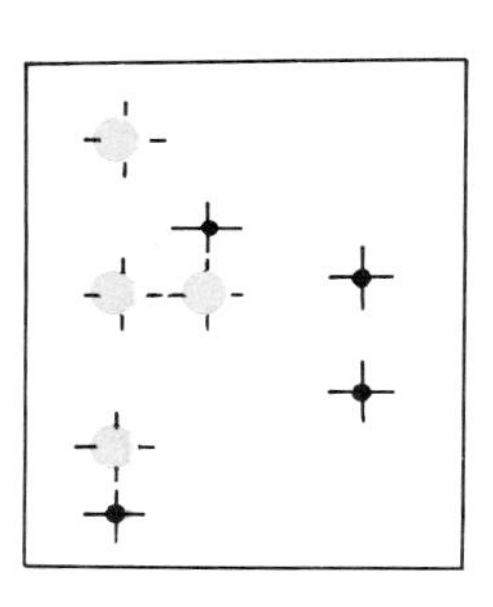

4 TAKE OFF THE TRACING PAPER AND PENCIL IN CROSSES TO HELP YOU CENTER THE SPOTS. STICK ON THE SPOTS.

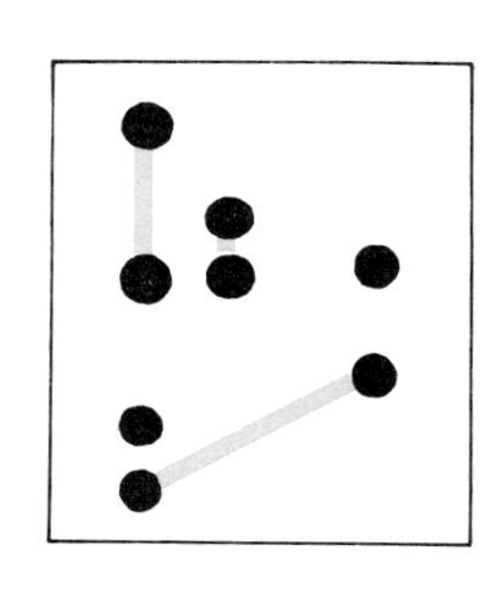

5 LOOK AT THE BOOK AND CONNECT THE SPOTS TOGETHER WITH TAPE.

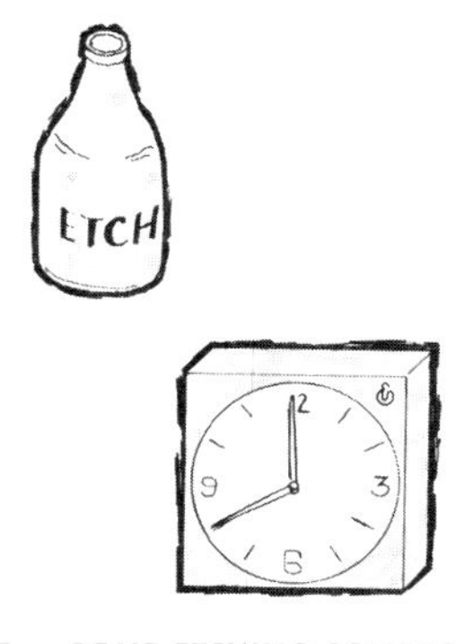

6 SOME ETCHING SOLUTION AND AN HOUR OR SO AND THE BOARD IS DONE.

7 CLEAN OFF THE SPOTS AND TAPE AND POLISH WITH STEEL WOOL.

Figure 6-2: These are the steps for making a printed circuit board.

SELF CHECK

1. Why is it necessary to go around the tape joints with a dull pencil?
2. What should you do before etching your layout?
3. What size tray should be used for the echant solution?
4. Why is it important to wear rubber gloves?

A tester is an essential tool for working on electrical of electronics projects. It also makes a great gift for the do-it-yourselfer or someone who works on cars.

A tester allows you to check electrical and electronic parts before you hook them up in a circuit. Even new parts can be bad.

The circuit board in Unit 6 is for a memory and switching amplifier module. This module can be used for a variety of projects. One use is in a tester.

If you do not have the materials to make printed circuit boards, you will find details in the Project Section on how to build the projects in this book without printed circuit boards.

The parts needed to build the tester are:

- 2 Light Emitting Diodes (LED)
- 1 Silicon Controlled Rectifier (SCR)
- 3 Resistors

The Silicon Controlled Rectifier (SCR) is the heart of the circuit. It is a very special kind of transistor.

The SCR is an electronic switch. The gate element can turn the SCR switch on, but it cannot turn it off. Once the SCR has been turned on it stays on. It "remembers" that it was turned on. The power must be turned off for a second to make the SCR turn off. We must have a "reset" button to erase the memory.

While every electrical and electronic part has a special symbol called its "schematic symbol", parts with the same symbol may not look alike. For example, SCRs come in several styles. Two of the most popular styles, and the schematic symbol for an SCR, are shown in figure 7-1.

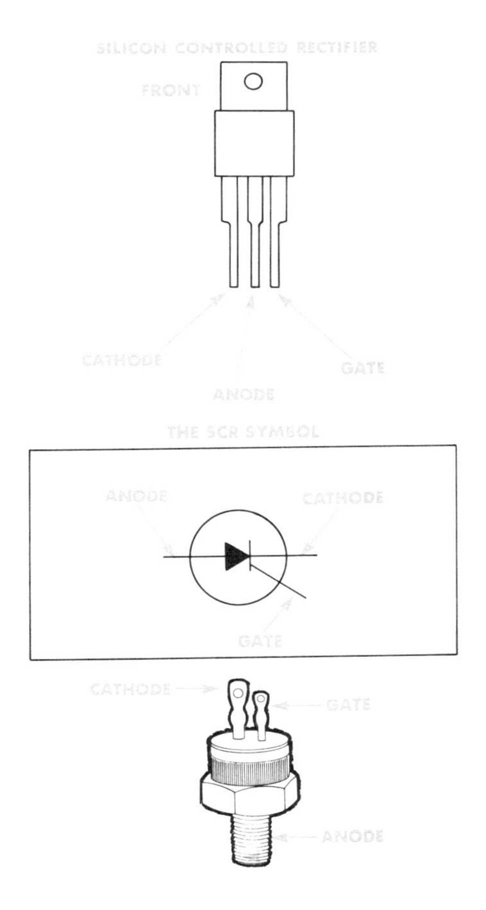

Figure 7-1: The Silicon Control Rectifier (SCR) is the heart of the tester circuit.

The tester module uses 2 Light Emitting Diodes (figure 7-2). A diode is an electronic valve that lets electrical current pass through it in one direction but not in the other direction.

The Light Emitting Diode (LED) is a special diode that produces light when electricity passes through it. One red and one green LED may be used in the tester, but two red ones will work just as well. Red LED's are easier to find. Also, yellow LED's can be used.

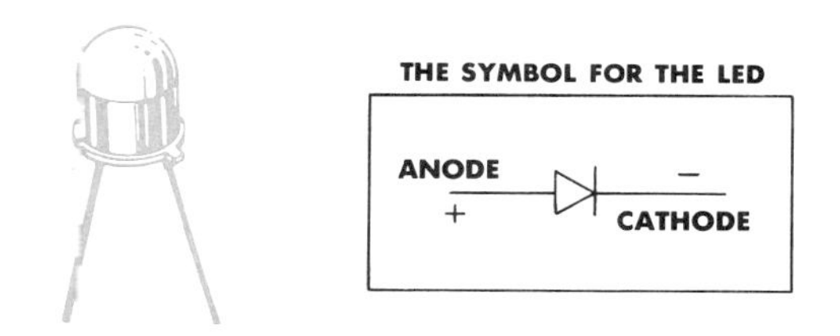

Figure 7-2: The Light Emitting Diode (LED) is a special diode that produces light when electricity passes through it. The anode and cathode are usually not marked.

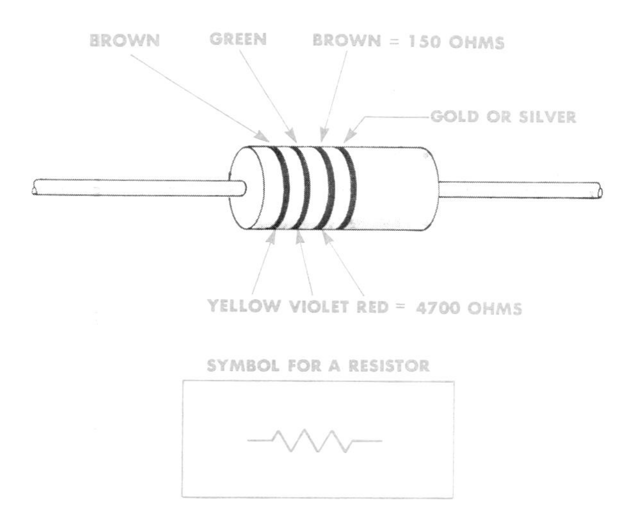

Figure 7-3: The value of a resistor is marked by color bands.

There are three resistors in the module. These resistors are used to keep the SCR and LED's from being overloaded and damaged.

The value of a resistor is marked by color bands (figure 7-3). Later you will learn how the color code works.

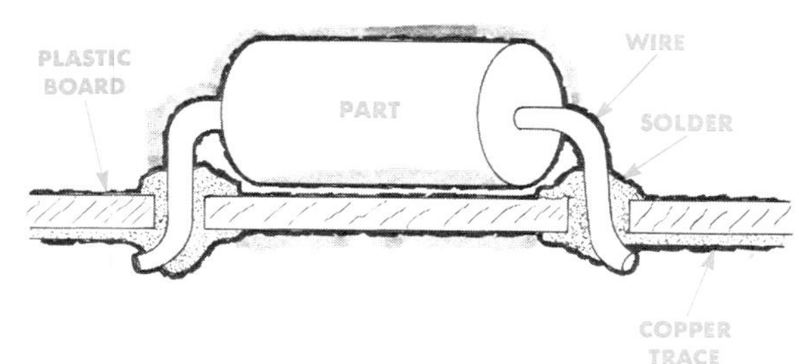

Figure 7-4: Soldering Parts on the P.C. Board.

1. Bend the leads (leeds) and push them through the holes in the circuit board. The wires on electronic parts are usually called leads.
2. Bend the leads slightly to keep the part from falling out before it is soldered in.
3. Solder the leads to the circuit board pads with a 25 to 35 watt soldering iron. Be careful not to keep the heat on too long or the copper may peel up off the board.
4. After the part has been soldered in, cut off the extra wire sticking out of the board.

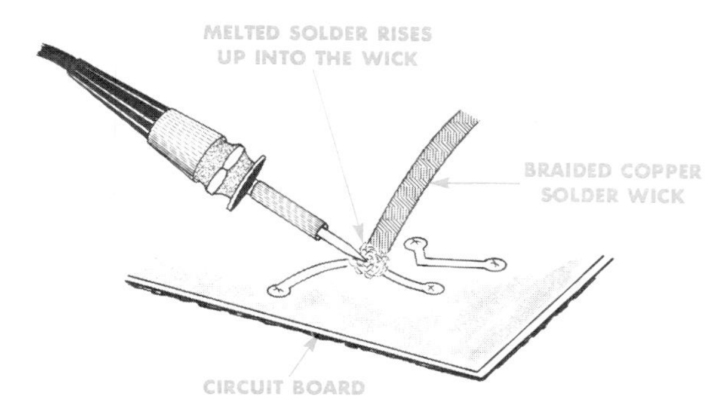

Figure 7-5: If you must unsolder parts from a circuit board, this figure shows you how to do it.

The unit name for resistance is Ohms. Resistors are ordered by so many Ohms just as a car travels so many miles.

To build the tester, first mount and solder the parts to the printed circuit (p.c.) board (figure 7-4). Bend the leads and push them through the holes in the circuit board. The wires on parts are usually called leads.

Bend the leads slightly to keep the part from falling out.

Solder the leads to the circuit board pads with a 25 to 35 watt soldering iron. Use only special solder intended for electronics work. Never use acid core solder. Be careful not to keep the heat on too long or the copper may peel up off the board.

After a part has been soldered in position cut off the extra wire sticking out of the board. Use diagonal cutting pliers.

If you must unsolder parts from a circuit board, figure 7-5 shows you how.

In figure 7-6, the symbols for all the parts of the tester have been connected together into a complete schematic diagram. Figure 7-7 shows how the parts go on the p.c. board. Both drawings say the same thing. Can you follow them through point by point to see how?

Before soldering the parts on the board, find out which wire on each LED is anode and which is cathode. LED's are not always marked. Figure 7-8

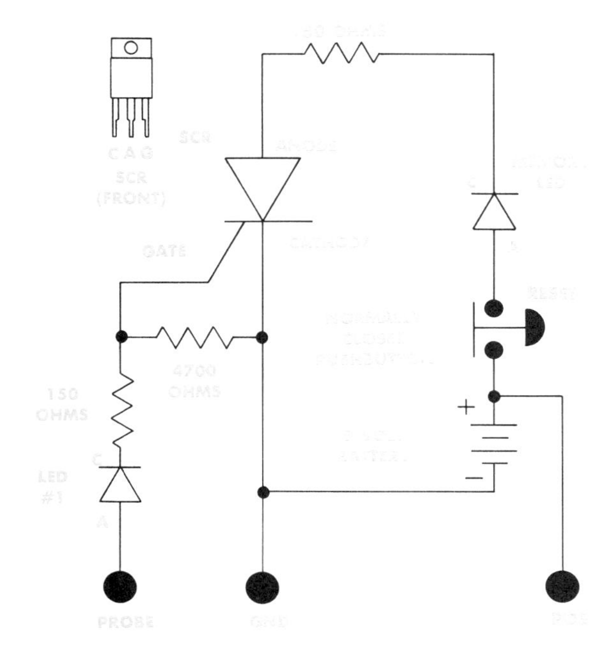

Figure 7-6: Schematic diagram of the tester.

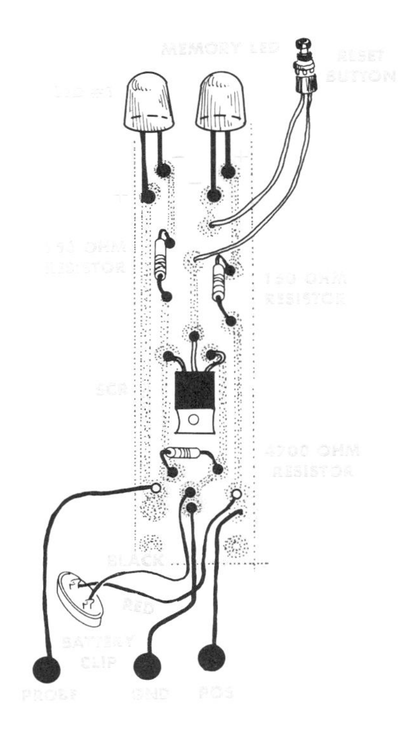

Figure 7-7: How the parts go on the tester circuit board.

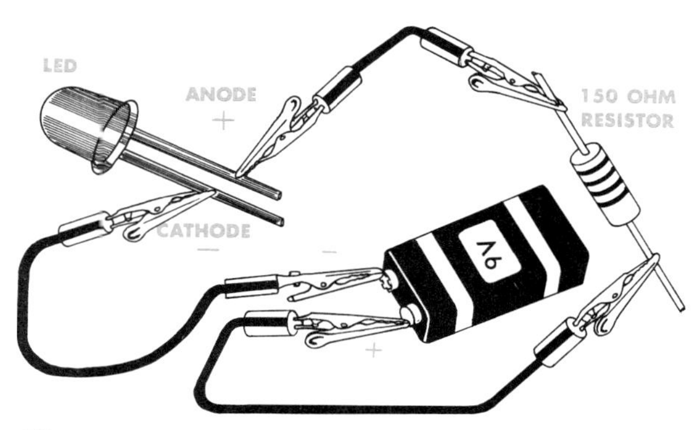

Figure 7-8: Checking the polarity of a LED. An LED will not work if it is installed wrong.

shows how to find out which lead is which. After finding out, put a piece of tape on the anode lead.

Now, go ahead and insert the parts into the board and solder them. Clip the extra wire sticking out of the board.

Finish the assembly as shown in figure 7-7. The tester can now be tested (figure 7-9).

If the tester doesn't work in Test A, the test steps in figure 7-10, Test B, should be taken.

There are two possibilities if the problem still hasn't been solved by Test B. The SCR can be bad or hooked up wrong, or LED #1 can be bad or hooked up backwards.

Make sure the clip lead between the cathode and the anode is disconnected before trying one more test (figure 7-11).

AUTHOR'S NOTE: I have never seen a new resistor that was bad. But watch out! It is very easy to read the color code wrong and put in a resistor with the wrong value. Even the professionals make this mistake.

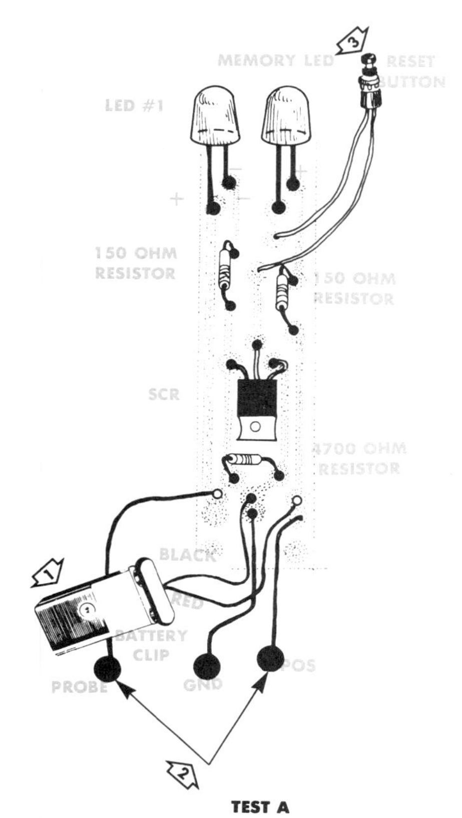

Figure 7-9: Troubleshooting the tester.

SELF CHECK

1. Why is it a good idea to test parts before installing them in a circuit?
2. Why should all LED's be tested before they are installed in a circuit?

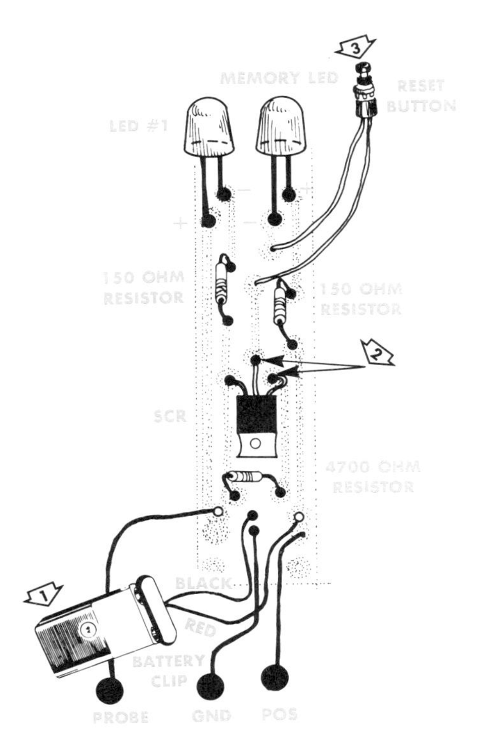

TEST B

1. **CHECK YOUR WIRING AND TRY A FRESH BATTERY. BE SURE TO CHECK THE COLOR CODE ON ALL RESISTORS!**
2. **CONNECT A CLIP LEAD BETWEEN THE SCR ANODE AND THE SCR CATHODE. THE MEMORY LED SHOULD LIGHT. IF IT LIGHTS, GO ON TO STEP 6.**
3. **IF THE MEMORY LED DOESN'T LIGHT PUSH THE RESET BUTTON. IF THE LED LIGHTS THEN YOU HAVE THE WRONG KIND OF PUSH BUTTON OR IT IS CONNECTED WRONG.**
4. **IF PUSHING THE BUTTON DOESN'T LIGHT THE LED, THE LED IS BAD OR YOU HAVE PUT IT IN BACKWARDS.**
5. **IF YOU HAVE FOUND THE TROUBLE, DISCONNECT THE CLIP LEAD GO BACK TO STEPS 3, 4, AND 5 IN TEST A.**

Figure 7-10: Troubleshooting the tester.

3. Match these:
   a. Anode 1. negative
   b. cathode 2. positive
4. Can you be sure that a new part will always be good?

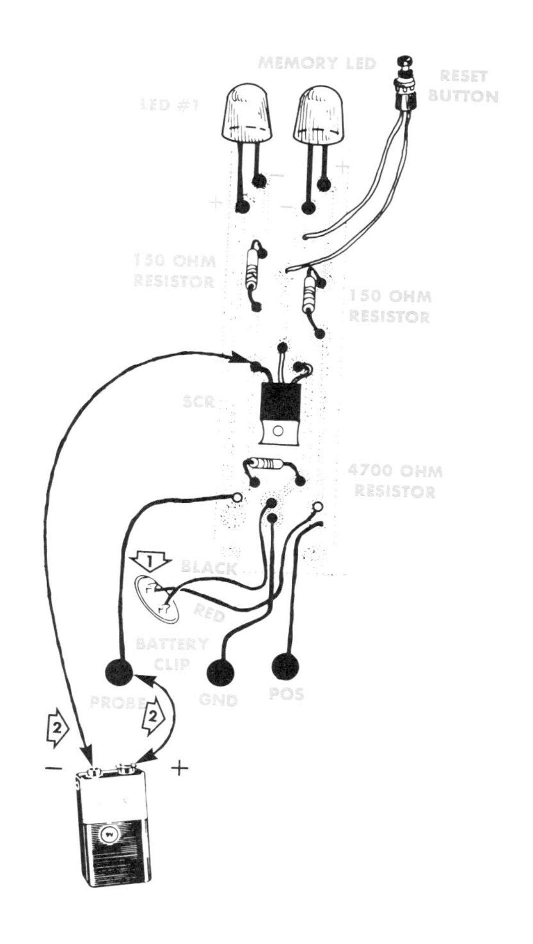

TEST C

1. **DISCONNECT THE BATTERY FROM THE CIRCUIT.**
2. **THEN CONNECT A CLIP LEAD TO EACH TERMINAL OF THE BATTERY.**
3. **CONNECT A CLIP LEAD FROM THE POSITIVE TERMINAL OF THE BATTERY TO THE PROBE INPUT. CONNECT THE CLIP LEAD FROM THE NEGATIVE BATTERY TERMINAL TO THE SCR GATE. IF THE #1 LED LIGHTS, THE PROBLEM IS THE SCR, REPLACE IT. IF LED #1 DOES NOT LIGHT, IT IS BAD OR IN BACK-WARD.**
4. **EXCHANGE THE CLIP LEADS. THE POSITIVE TERMINAL OF THE BATTERY WILL NOW BE CONNECTED TO THE SCR GATE AND THE NEGATIVE TERMINAL TO THE PROBE. IF LED #1 LIGHTS NOW, YOU HAVE IT IN BACKWARD. IF IT DOESN'T LIGHT IT IS BAD.**

Figure 7-11: Troubleshooting the tester.

Unit 8 TESTING ELECTRONIC PARTS

Electronic parts are usually called "components". Testing components is one way to find out why a project doesn't work right. One way to test a project without a tester is to turn it on and watch it burn up. This is called "the smoke test". It is not recommended (figure 8-1).

To practice testing components, you will need one of each of the following components:

- Fuse
- Switch or push button (any type)

Figure 8-1: The smoke test is not the best way to test electronic components.

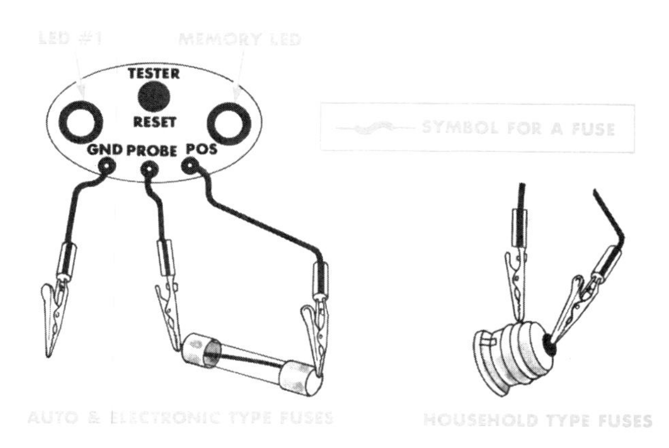

Figure 8-2: A fuse is OK if: LED #1 is lit, memory LED is lit, and the tester won't reset with the fuse connected.

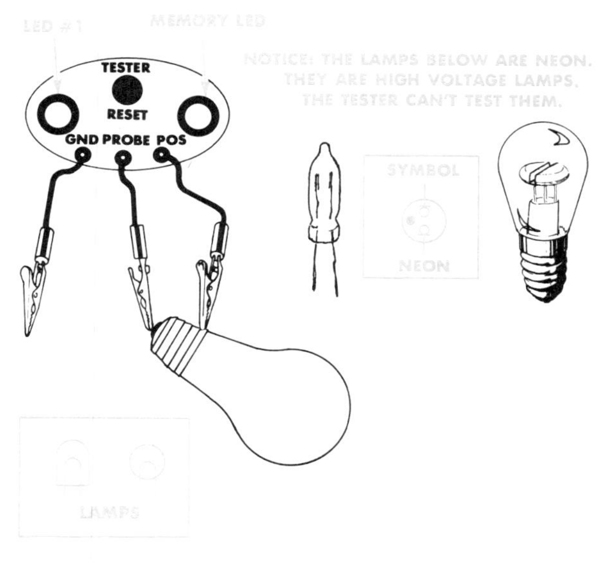

Figure 8-3: A lamp is OK if: LED #1 is lit, memory LED is lit, and the tester won't reset with the lamp connected. NOTE: Lamp being tested will not light.

- Resistor 150 ohms (black-green-brown)
- Household light bulb or flashlight bulb
- Capacitor, any value 50 microfarads (50uF) or larger
- Capacitor, any value from .047 to 0.5 microfarad (.047uF to .5uF)
- Silicon diode (any type)
- LED (any color)
- NPN transistor (any type number)
- PNP transistor (any type number)
- SCR (any kind)

Figures 8-2 through 8-8 show how to test the components with the tester you built in Unit 7.

After you finish this unit and learn the resistor color code in the next unit, you know enough to build any of the projects in the project section in the book.

SELF CHECK

1. What does POS on the tester mean?
2. What does GND on the tester mean?
3. What does the reset button on the tester do?
4. Do you know another name for electronic parts?

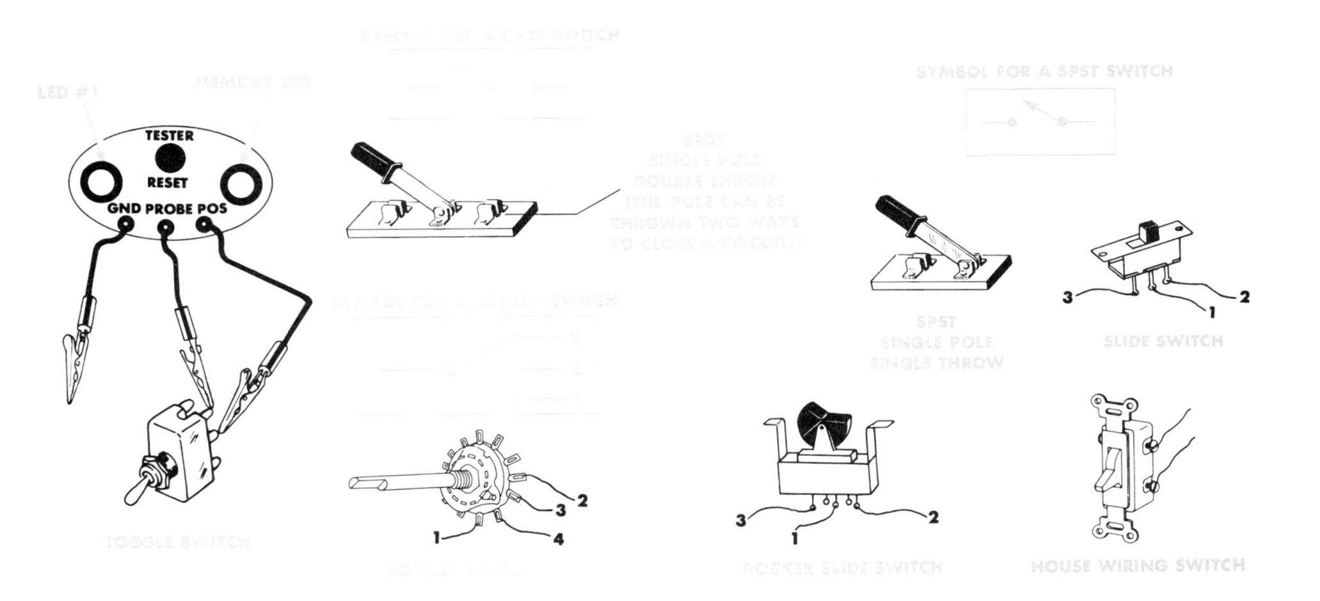

Figure 8-4: To test a switch, start with the switch on. If the switch is off both LEDs will stay dark.

1. Switch <u>on</u>—LED #1 is lit, memory LED is lit, and the tester won't reset with the switch on.
2. Switch <u>off</u>—LED #1 goes out, memory LED stays on, and memory LED will reset.

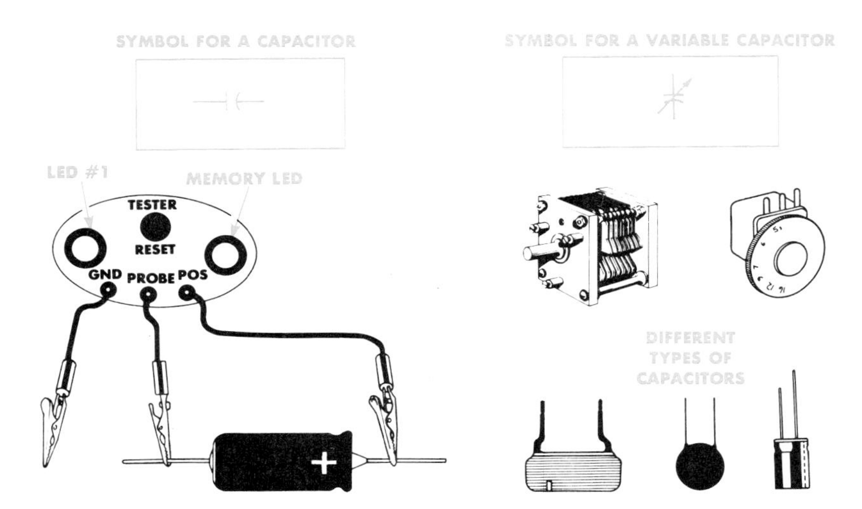

Figure 8-5: A capacitor is OK if: LED #1 flashes and goes out, the memory LED stays lit, and the memory LED will reset with the capacitor connected. If it doesn't the capacitor is shorted. NOTE: The tester will not give a complete test for capacitors smaller than .033 microfarad, but the tester will indicate a shorted capacitor of any value. A capacitor is shorted if: LED #1 is lit, memory LED is lit, and the tester will not reset.

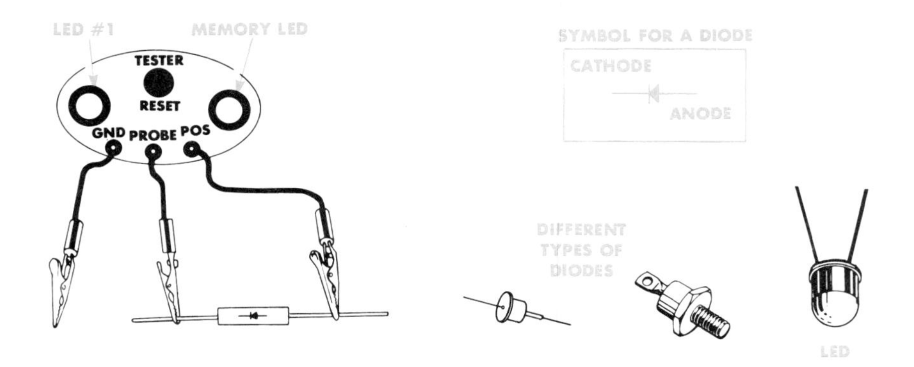

Figure 8-6: A standard diode or a LED is OK if:

1. With pos. connected to the anode and probe connected to the cathode—LED #1 is lit, memory LED is lit, and the tester won't reset. NOTE: A LED being tested will light.
2. With pos. connected to the cathode and probe connected to the anode:
   a. Both LED's are dark.
   b. A slight glow in LED #1 means you have a leaky diode—throw it away.
   c. Memory LED lit means a shorted diode—throw it away.

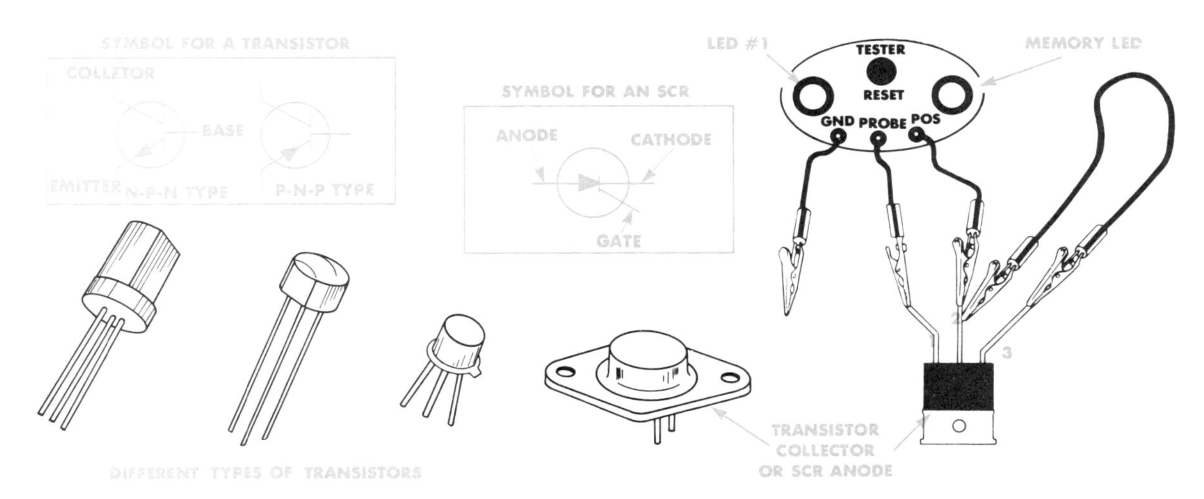

Figure 8-7: Testing transistors.

1. Transistor connections are not standard and it is necessary to find out which lead is Base, which is Emitter, and which is Collector. The transistors shown here are just a few of the types you will run into. You can look them up in a transistor handbook, or ask your instructor. Basically, there are only two types of transistors PNP and NPN. For example, the transistor connected to the tester in this drawing is an NPN type. #1 is Emitter, #2 is Collector, and #3 is base. To test a PNP type exchange the probe and pos. leads of the tester. For all SCRs: #1 is cathode, #2 is anode, and #3 is gate.
2. Test Procedure (for transistors and SCRs):
   a. Always connect the leads from the tester first. If LED #1 glows the transistor or SCR is leaky. If the memory LED lights, the transistor or SCR is shorted.
   b. Now connect the clip lead from collector to base for a transistor; or from anode to gate for an SCR. Both LED's should light. If they don't the transistor or SCR is bad.

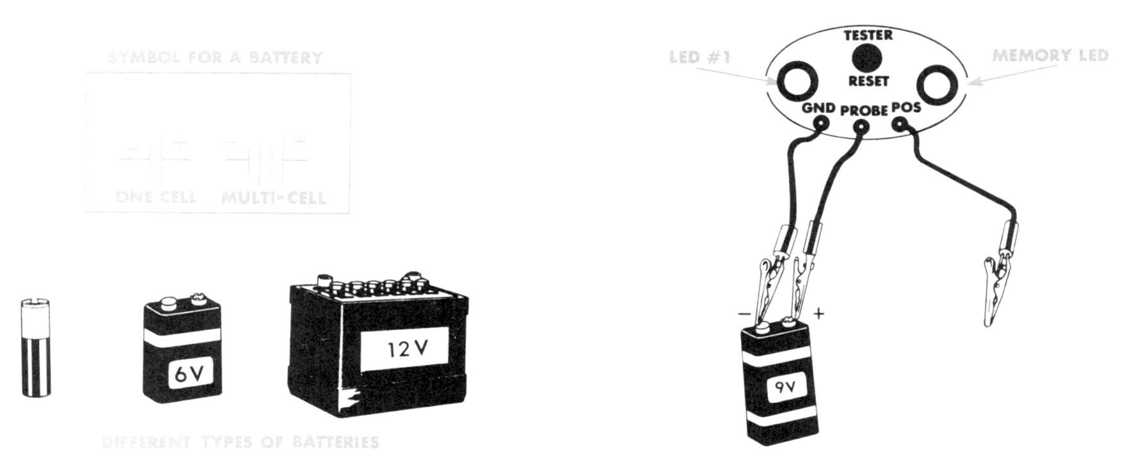

Figure 8-8: A Battery is OK if: both LED's light and the tester cannot be reset with the battery connected.
NOTE: The lowest voltage the tester will test is about 5 volts.

Unit 9 BREAKING THE RESISTOR COLOR CODE

The value of carbon resistors in not printed on them in numbers. Instead, the value is in code in the form of colored bands (figure 9-1). Each color stands for a digit between 0 and 9.

Figure 9-2 is a color code decoder sheet. If you read the last column from the top down, you will find a message that will help you remember the code. Each word in the message starts with the same letter as the color listed to the left of it.

There are little cardboard color wheels that you can get to figure the code out for you. DON'T GET ONE! Memorize the color code and it will always be with you. Buy a color wheel and you will spend half your shop time looking for it. If you think a color wheel will help you learn the code, you are WRONG. The human brain doesn't

DIGIT	COLOR	MEMORY HELPER
0	BLACK	BIG
1	BROWN	BOYS
2	RED	RACE
3	ORANGE	OUR
4	YELLOW	YOUNG
5	GREEN	GIRLS
6	BLUE	BUT
7	VIOLET	VIOLET
8	GRAY	GRAY
9	WHITE	WINS

Figure 9-2: Color code decoder sheet.

Figure 9-1: Carbon resistors are color coded.

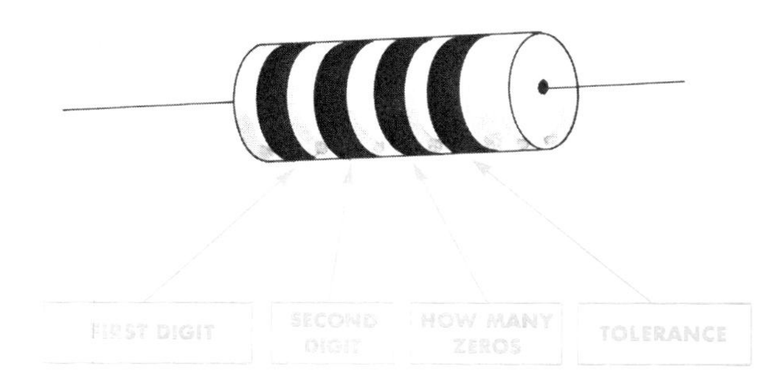

Figure 9-3: Resistor color bands are read from left to right.

work that way. It won't take long to learn it, so go ahead and do it.

The value of the resistor is read from left to right. Start with the color band closer to one end of the resistor. This is the first number. Write down the number that goes with the color. The second band is the second number. Write down the number that goes with that band's color. The third band is the number of zeros. Write down that many zeros. Some examples are shown in figure 9-4.

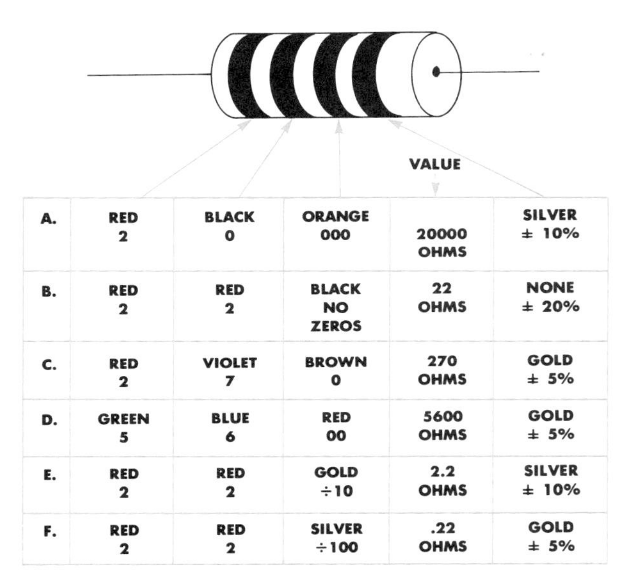

				VALUE	
A.	RED 2	BLACK 0	ORANGE 000	20000 OHMS	SILVER ± 10%
B.	RED 2	RED 2	BLACK NO ZEROS	22 OHMS	NONE ± 20%
C.	RED 2	VIOLET 7	BROWN 0	270 OHMS	GOLD ± 5%
D.	GREEN 5	BLUE 6	RED 00	5600 OHMS	GOLD ± 5%
E.	RED 2	RED 2	GOLD ÷10	2.2 OHMS	SILVER ± 10%
F.	RED 2	RED 2	SILVER ÷100	.22 OHMS	GOLD ± 5%

Figure 9-4: Different colored bands equal different resistance values.

The unit for resistance is the Ohm. The greek letter Ω is often used instead of writing out the word "Ohm".

Resistors often have a fourth band that tells tolerance (how far off the resistor might be).

Tolerance is rated in percent. Percent is another way of saying: "How many cents on the dollar?"

Suppose somebody hands you a sack of pennies. The person tells you they have not been counted. They have been weighed, and there is $1.00 (100 cents) plus or minus 10% (10 cents) in the bag. This means that the bag could contain anywhere from 90 cents to $1.10 (figure 9-5).

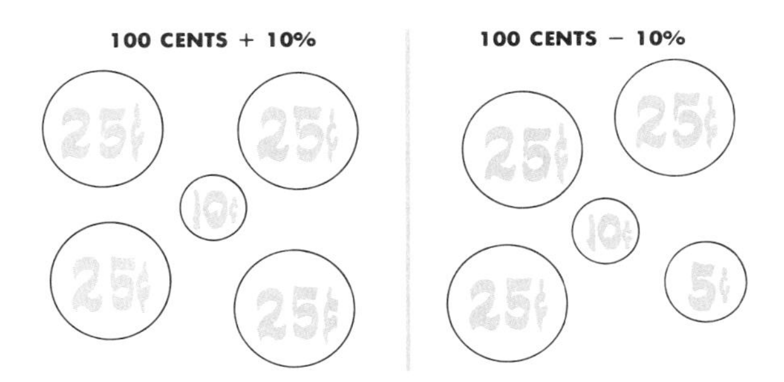

Figure 9-5: Tolerance is rated in percent. Percent is another way of saying "How many cents to the dollar".

In the case of a resistor a 100 Ohm (plus or minus 10%) resistor could have any value between 90 Ohms and 110 Ohms.

Some resistors have a fifth band. The fifth band is nearly always yellow. The fifth band is a military requirement and is of no importance to anyone else.

A resistor converts electrical energy into heat. The more heat the resistor must get rid of, the bigger it must be. The small resistors in figure 9-6 are color coded. The power resistors shown in figure 9-6 are much larger. The value of the bigger resistors in figure 9-6 are printed on the body. They are not color coded.

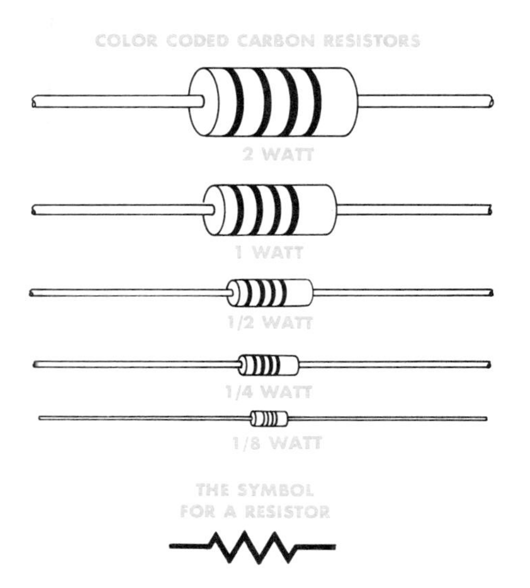

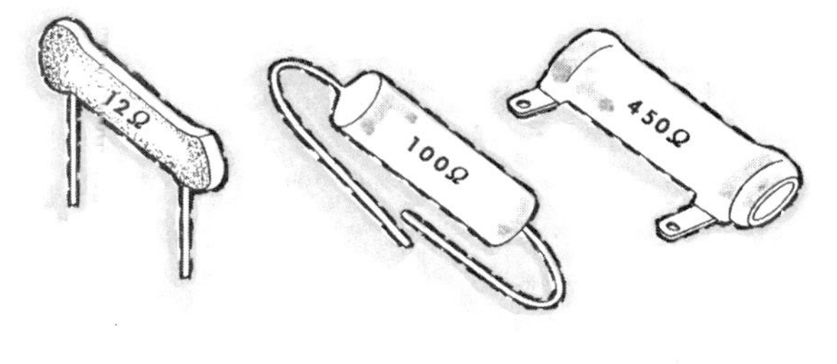

Figure 9-6: Resistors convert electrical energy into heat. The More heat the resistor must get rid of, the bigger it must be.

Variable resistors come in a number of shapes and sizes. Variable resistors are often called potentiometers, or simply pots. Some pots have the value stamped on them. Unfortunately many of them have a bunch of useless code numbers on them, and nothing else.

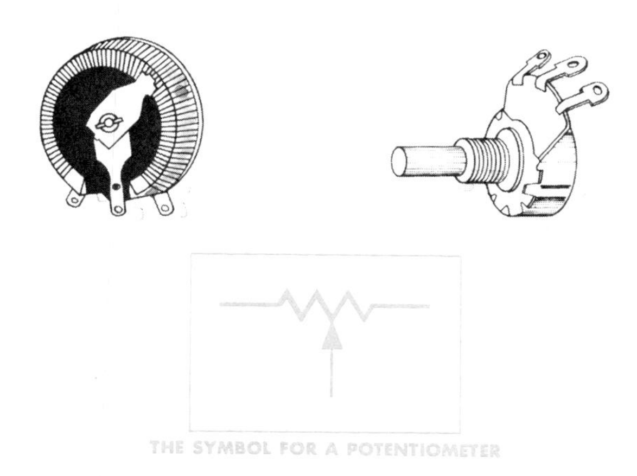

Figure 9-7: A sliding contact on variable resistors allows us to change their value.

When you buy a potentiometer look at it as soon as you open the package. If it isn't marked, mark it.

SELF CHECK

Here are some color code samples. See if you can find the value in Ohms for each one.

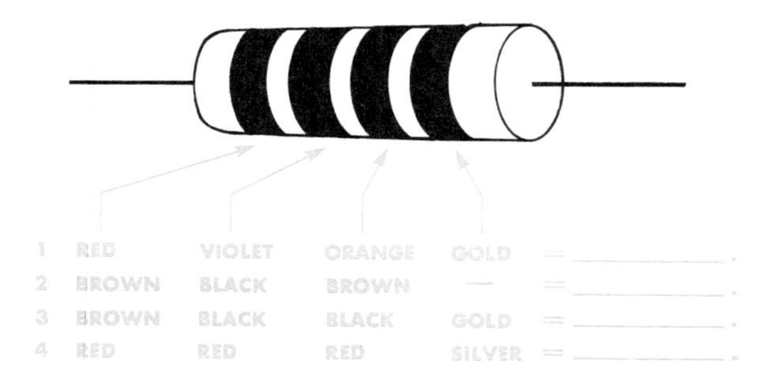

1	RED	VIOLET	ORANGE	GOLD	= ______.
2	BROWN	BLACK	BROWN	—	= ______.
3	BROWN	BLACK	BLACK	GOLD	= ______.
4	RED	RED	RED	SILVER	= ______.

Electrons flowing through a circuit can make things work. Electricity is nothing more than the flow of electrons. These "flowing" electrons come from the outside ring (orbit) of metal atoms, and a few other materials that can carry electricity.

Materials with electrons that can leave their atoms easily are called "conductors". Most metals are good conductors because the electrons in their outer orbit are held to the nucleus by very weak forces (figure 10-1). These electrons can move through the metal easily.

Materials which cannot carry electricity are called non-conductors or "insulators". Insulators are made of non-metallic substances such as glass, wood, and most plastics. They have electrons that are held to the nucleus by very strong forces (figure 10-2).

Semiconductors are not very good conductors and not very good insulators. Their electrons are tied to the nucleus by medium forces. They are not always free to travel like the electrons in metal "conductors". They are not held so tightly that they can't travel at all like those in "insulators".

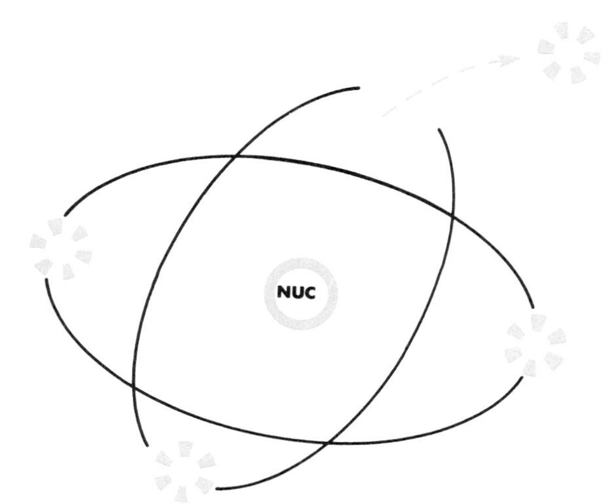

Figure 10-1: The electrons in a conductor are free to leave the atom.

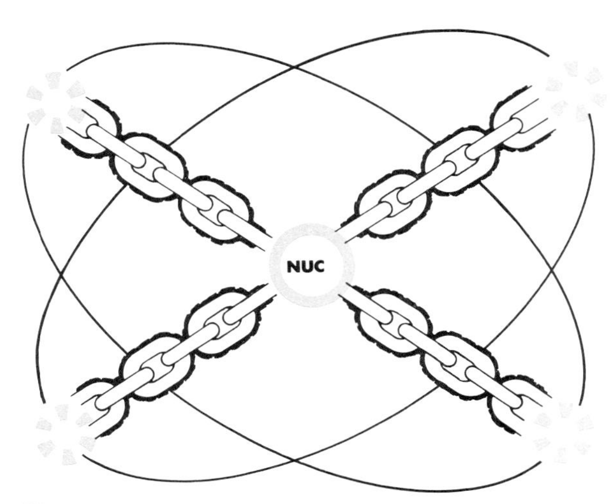

Figure 10-2: The electrons in an insulator are held in place by strong forces. They are not free to leave the atom.

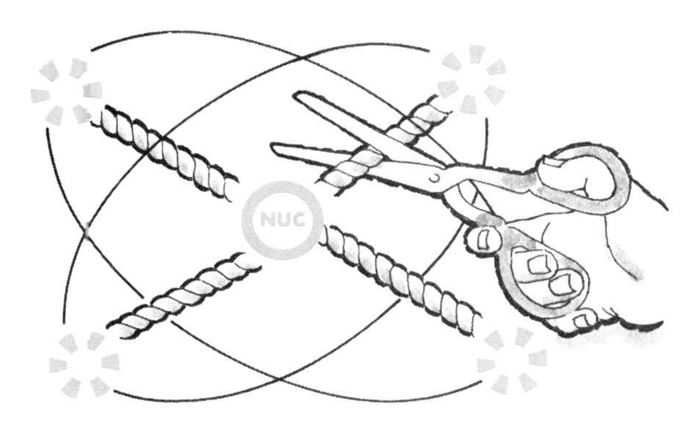

Figure 10-3: The electrons in a semiconductor are not free to leave the atom unless they get outside help from heat or light.

However, with the help of an outside energy such as heat or light, the electrons in semiconductors can get free to travel (figure 10-3).

There are only a few semiconductor materials. The most important are: silicon, carbon, and gallium arsenide. Nearly all transistors and integrated circuits are made of silicon. Pure silicon looks almost like metal but is brittle like glass. Every time you go to the beach or desert you carry home enough silicon in your clothes to make a batch of transistors. Sand is composed mostly of silicon and oxygen, so there is plenty of silicon around.

Light emitting diodes are made from gallium arsenide.

Carbon is used to make batteries and resistors. Most color coded resistors are made of carbon.

Liquid can be an insulator or conductor. Pure water is an insulator. But we hardly ever find pure water. A trace of salt turns water into a good conductor. Since the human body is mostly water and salts, it is a good conductor (figure 10-4).

Figure 10-4: Electricity is dangerous because the human body is a good conductor.

On the other hand, a little contamination doesn't turn a conducting liquid into a non-conductor.

SELF CHECK

1. Electricity flows best through:
   a. insulators
   b. semiconductors
   c. conductors?
2. What material has "loose" electrons?
3. What is the most common semiconductor material?
4. Is pure water a conductor or an insulator?

Electricity is moving electrons. It takes energy (called voltage) to push electrons (current) through wires and electrical devices. Electrons in motion always encounter some resistance. Resistance is electrical friction.

The three parts of electricity (figure 11-1) are:

- Current—the movement of electrons
- Voltage—the force needed to move the electrons
- Resistance—electrical friction that opposes the movement of electrons

When water is flowing in a stream we often talk about the current in the stream. We also use the word current when we talk about electrons flowing in an electrical circuit. The amount of current in an electrical circuit depends on how many electrons are flowing. Many electrons flowing is a large current and few electrons is a small current.

Electrical current is measured in amperes (amps). It takes 6,250,000,000,000,000,000 electrons flowing past a single point in a wire each second to make 1 amp of current (figure 11-2).

Figure 11-1: The three parts of electricity.

Figure 11-2: 6,250,000,000,000,000,000 electrons are called one COULOMB. The word Coulomb is pronounced "Cool-ohm" but is has nothing to do with resistance or ohms.

Figure 11-3: Voltage is the energy it takes to push electrons through an electrical circuit.

Figure 11-4: Electrons are shoveled into a trough until it overflows at the positive end. Moving anything, including electrons, takes work. Work means that some kind of energy is being used. In this case it is muscle energy.

Voltage is the energy it takes to push electrons through an electrical circuit. Voltage is created by piling more electrons on one end of a wire than there are on the other end (figure 11-3).

The end of the "wire" with the highest pile of electrons is called the negative end because there are more negatively charged electrons at that end. The end of the wire with fewer electrons is called the positive end. Electrons flow from negative to positive.

Actually both ends of the wire are negative and one end is just less negative than the other, but it is called positive anyway. At one time it was believed that electricity was made up of two "fluids". One "fluid" was believed to be positive and one negative. One end of a circuit has been called positive and one end negative ever since.

Electrons will stop flowing when they are spread evenly along the path of their flow. In other words, one end of the trough must have electrons piled higher than the other to start the flow again (figure 11-4).

To get a continuous flow of electricity, there must be a steady input of energy (figure 11-5). To get this input, all of the electrons must be recycled. When the electrons are recirculated we have a complete circuit. Electrons can light the room, turn motors, and do other work as they flow around the circuit.

As electrons move through an electrical circuit they rub against other electrons and atoms. Like any other kind of friction, this rubbing produces

Figure 11-5: A steady input of energy is required to get a continuous flow of electricity.

Figure 11-6: Any kind of friction creates heat. Electrical friction is called resistance.

heat (figure 11-6). Some of the voltage in the circuit is converted into waste heat.

Friction in an electrical circuit is called resistance. The amount of resistance is measured in Ohms.

All conductors have some resistance. Platinum, gold, silver, copper, and aluminum have the least resistance of all the metals.

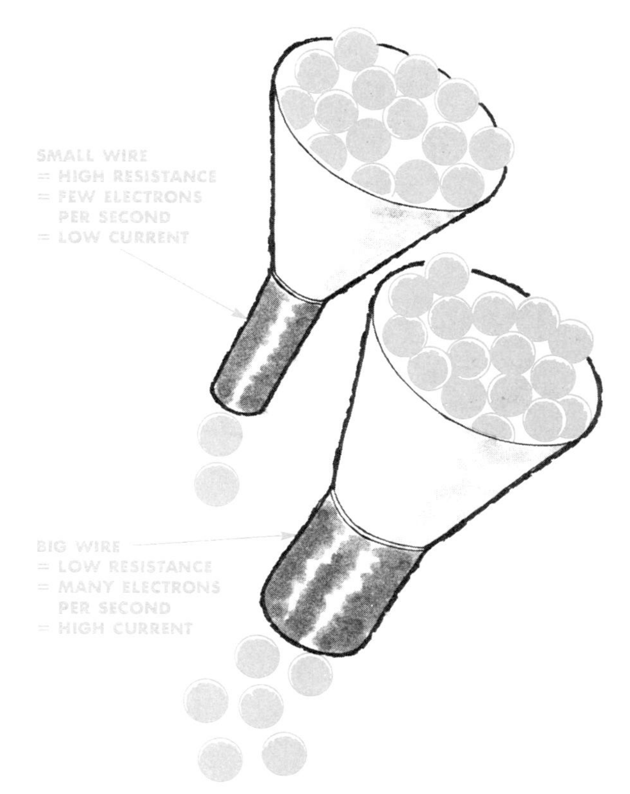

Figure 11-7: More current can flow through a big wire than a small wire.

Four rules of resistance are:

- The better conductor a material is, the less resistance it has.
- There is more resistance in a long wire than in a short one.
- A smaller diameter wire has more resistance than a wire with a bigger diameter (figure 11-7).
- Resistance reduces the amount of current that can flow through a circuit.

When working with an electronic circuit, each component has a certain current rating. If too much current is allowed to flow through any com-

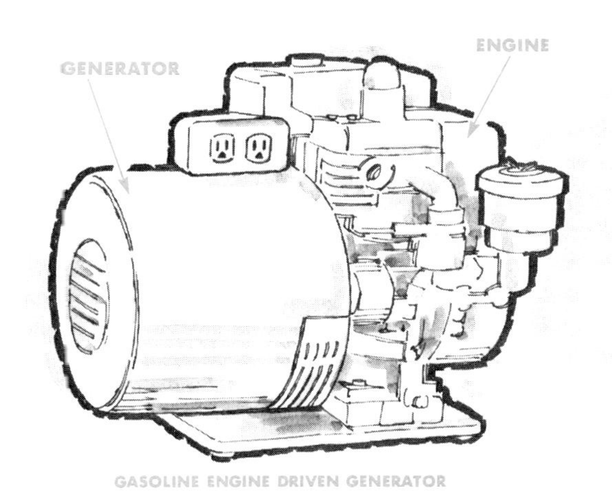

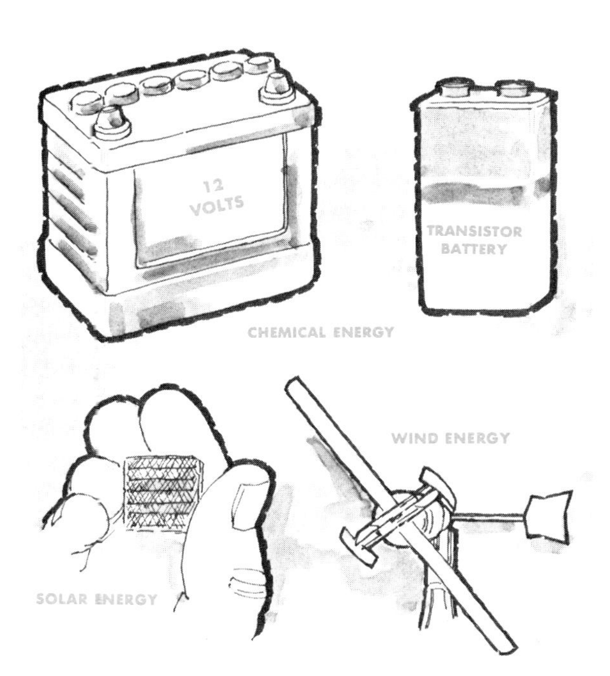

Figure 11-8: The energy needed to pile up electrons to start electricity flowing can come from many sources. Electricity can be generated by using mechanical energy, heat energy, light energy, chemical energy, and so on. Whatever way it is done, some kind of energy must be put in to get electrical energy out. We can't get something for nothing.

ponent, the component will overheat and destroy itself. A resistor of just the right value can be used to reduce the current passing through each component to a safe value.

For instance, you used a resistor to protect each Light Emitting Diode (LED) when you built the tester in Unit 7. A LED will absorb enough current to destroy itself if a resistor is not used to limit the current to a safe value.

Electrical energy cannot be created. Other forms of energy must be converted into electrical energy (figure 11-8). Gasoline is turned into mechanical energy by an engine which drives a generator. The generator motion (mechanical energy) moves electrons which can be measured as voltage. Coal or oil can be burned to make steam. The steam can turn a turbine that turns a generator.

Sunlight becomes electrical energy by using solar batteries. Hydroelectric generators, driven by water power, make electricity.

Chemical energy is converted into electrical energy in batteries.

SELF CHECK

1. Which end of the wire has the most electrons, positive or negative?
2. What is electrical current?
3. A certain kind of force is needed to push electrons through wire. What is the name of the force?
4. What is the name given to electrical friction?

A man named Professor Ohm figured out how voltage, current and resistance work together. He was forced to leave his job at the university because his fellow professors thought his idea was a bit crazy (figure 12-1). Now we know Professor Ohm was right on target. What he had to say is called Ohm's Law:

- One volt will push one ampere of current through one ohm of resistance

Professor Ohm also gave us a simple formula to find the value of one of the three parts if the other two are known. The formula for Ohm's Law can be shown in a magic triangle (figure 12-2).

Figure 12-1: Nobody believed in Ohm's law at first. Now, it's one of the most important laws of electronics.

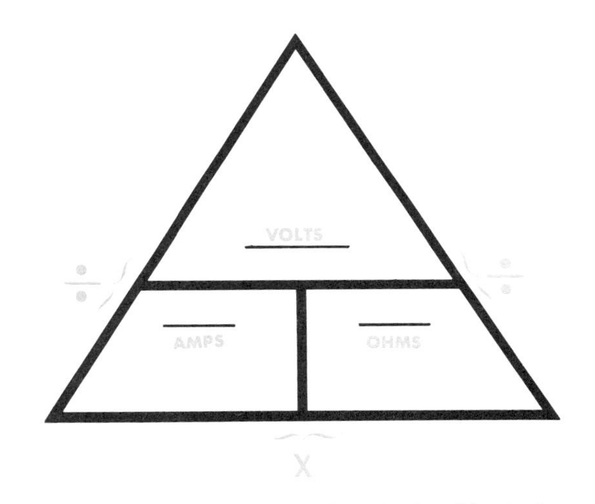

Figure 12-2: The magic triangle for Ohm's law.

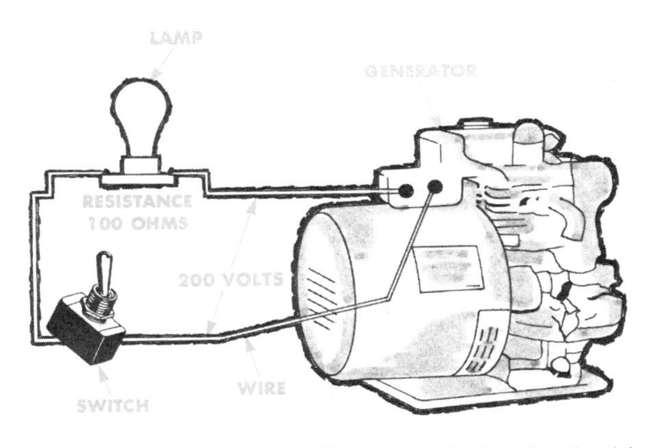

Figure 12-3: How much current is flowing in this circuit?

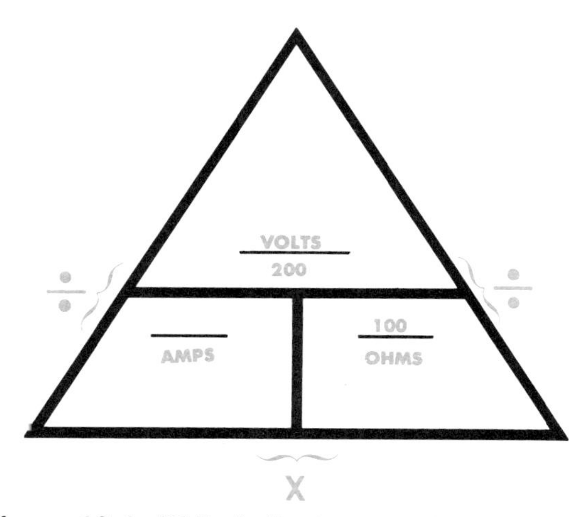

Figure 12-4: Write in the known values.

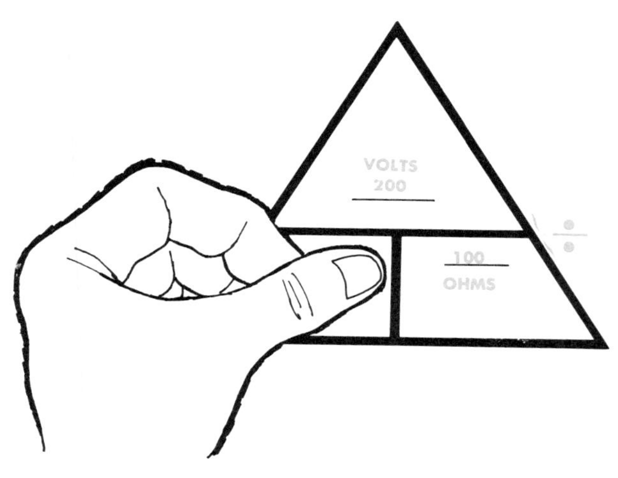

Figure 12-5: The magic triangle tells you to divide 200 volts by 100 Ohms. The answer is 2 amps.

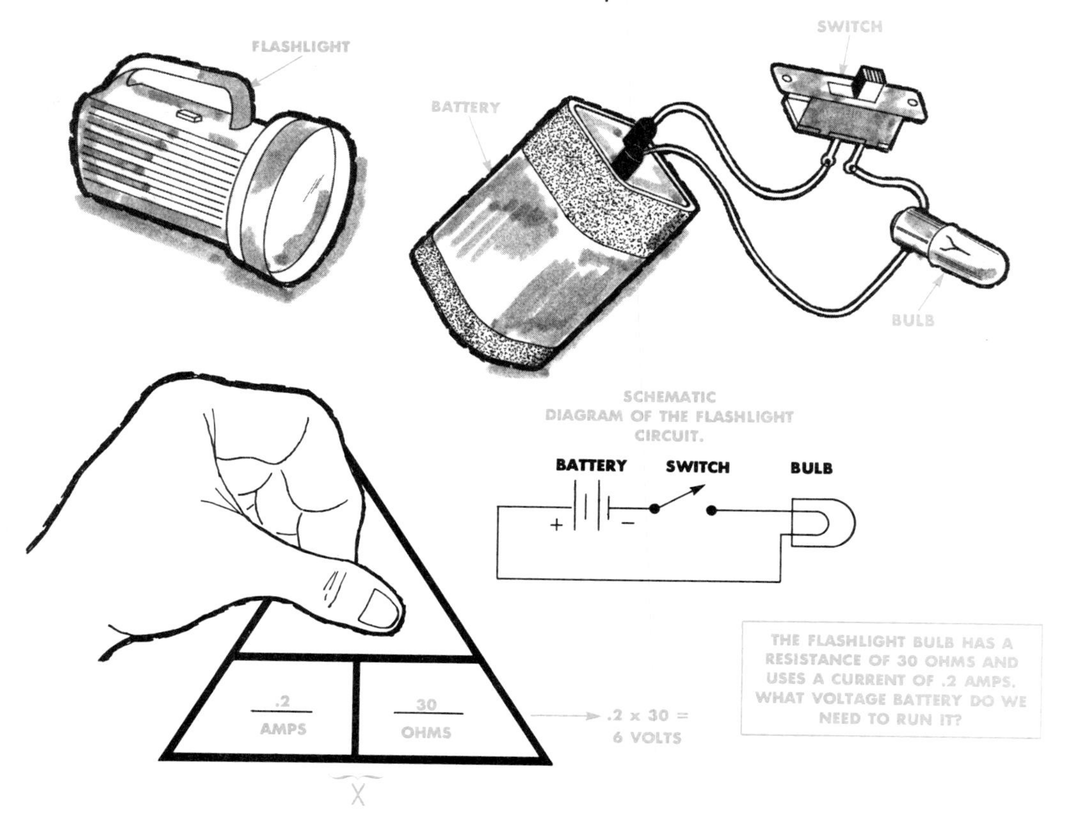

Figure 12-6: What voltage battery is needed to make this circuit work?

To make the magic triangle work, simply fill in what is known this way:

- Suppose the lamp in figure 12-3 has a resistance of 100 ohms, and the generator is producing 200 volts. How much current (how many amps) is flowing around the circuit?
- To solve the problem, write in what is known from figure 12-4:
  Volts = 200
  Ohms = 100
- What is not known is how many amps are flowing. To find the current (number of amps), cover the word AMPS on the triangle. The triangle will tell you what to do next (figure 12-5).

In figure 12-6, the resistance of the flashlight bulb is given in ohms and the current of the bulb is given in amperes. What is not known is the battery voltage. When the battery goes dead, what voltage battery should be bought to replace it? Fill in the numbers on the magic triangle and find the answer.

DO YOU KNOW

1. How many amps will an appliance use if it has 10 volts and 5 ohms?
2. How many ohms will an appliance have if it has 3 amps and 9 volts?
3. How many volts will an appliance need if it has 7 ohms and 2 amps?
4. How many ohms will an appliance have if it has 100 volts and 3 amps?

Unit 13 ELECTRICAL POWER

Electrical power is rated in watts or kilowatts. Kilo means one thousand, so a kilowatt is 1000 watts. Power in an electrical circuit is found by multiplying volts times amps. The formula is Watts = Volts × Amps, and we can make a "magic triangle" like Professor Ohm's to help us solve power problems (figure 13-1).

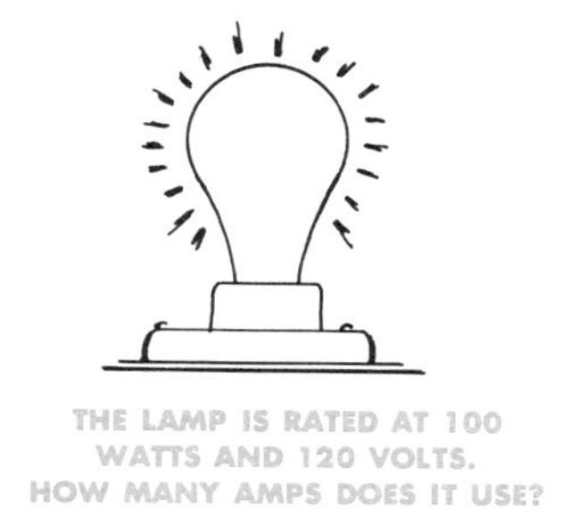

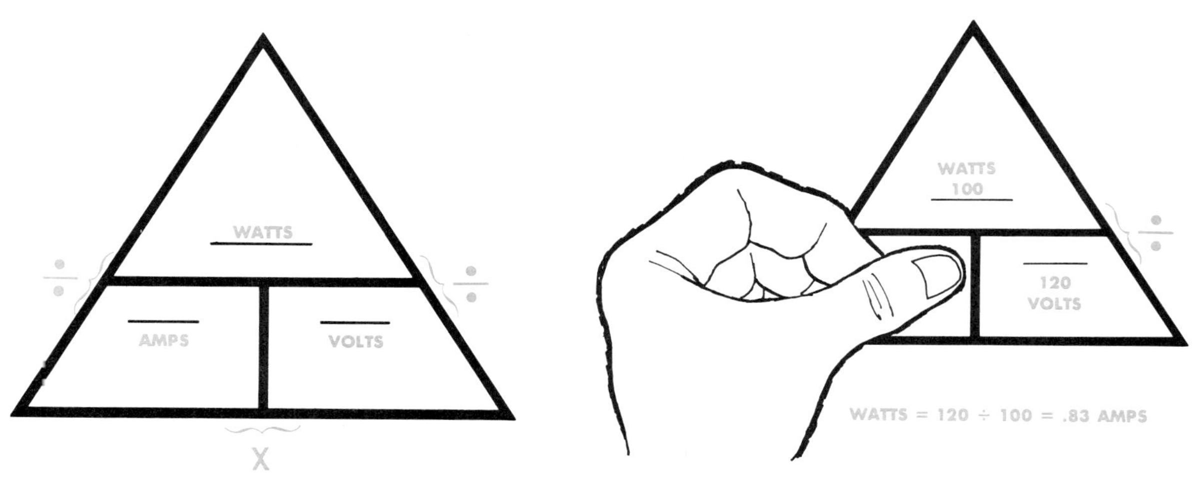

Figure 13-1: This is the magic triangle for power problems. Use it just like the magic triangle for Ohm's law.

DEVICE	POWER CONSUMED (WATTS)	EQUIVALENT HORSEPOWER
TELEVISION RECEIVER	50 TO 150	1/16 TO 2/10
RADIO RECEIVER	1 TO 50	1/1000 TO 1/16
ELECTRIC CLOCK	2	1/500
ULTRA-VIOLET LAMP	385	4/10
ELECTRIC FAN	100	1/10
PORTABLE ELECTRIC HEATER	1000	1 1/3
RECORD-CHANGER	75	1/10
ELECTRIC BLANKET	200	1/3
HEATING PAD	60	1/10
ELECTRIC SHAVER	12	1/200
REFRIGERATOR	150	1/5
COFFEE MAKER	1000	1 1/3
FLOOR LAMP	300	4/10
VACUUM CLEANER	125 TO 300	2/10
ELECTRIC SKILLET	1100	1 1/2
MIXER	100	1/10
TOASTER	1500	2
BROILER	1500	2
WAFFLE IRON	1000	1 1/3
WASHER	700	9/10
50 WATT STEREO AMPLIFIER	150	2/10
AUTOMATIVE HEADLIGHTS	250	3/10
HAND IRON	1000	1 1/3

Figure 13-2: The power used by a few typical electrical devices.

Figure 13-2 is a list of common electrical and electronic devices. It gives power consumption for each device. 746 watts equals 1 horsepower.

SELF CHECK

1. Which appliance listed in figure 13-2 uses the most power?
2. How many 100 watt light bulbs could operate with the power needed for a coffee maker?
3. Which would be cheaper, listening to your stereo for one hour or broiling a steak for 10 minutes?
4. How can you find the current of an automobile tail light bulb rated at 12 volts and 20 watts?

WAYS TO CONNECT BATTERIES

A battery is made up of one of more cells. A chemical reaction in each cell produces a set amount of electrical energy—from 1.2 volts to 2 volts depending on the kind of cell. A wet cell battery produces 2 volts per cell. A dry cell battery produces 1.5 volts per cell. It doesn't matter if it is a AA or a D size cell. Larger cells hold more electrical energy than smaller cells.

Cells are rated in ampere hours. For example, a 10 ampere hour cell will deliver 10 amperes for one hour, or 1 ampere for 10 hours, or 2 amperes for 5 hours.

One cell by itself is called a cell; two or more cells, connected together is called a battery. There are two ways to connect cells into batteries. They are:

- Parallel
- Series

Battery cells wired in parallel have more amp-hours than a single cell (figure 14-1). Three 1 amp-hour cells connected in parallel have the same energy capacity as a bigger 3 amp-hour cell, but the output voltage is still the same as for a single cell.

When several cells are connected in series, the cell voltages add up. A 9 volt transistor battery, for example, has six 1.5 volt cells hooked up in series inside the case. Three 1 amp-hour cells connected in series still provides 1 amp-hour capacity, but the output voltage is 4.5 volts (figure 14-2).

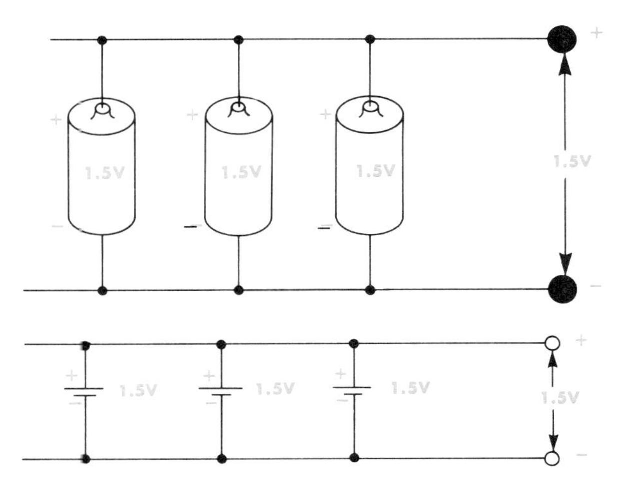

Figure 14-1: 3 batteries in parallel give the same voltage as 1 battery. But, 3 batteries in parallel provide 3 times the amper-hour capacity of a single battery.

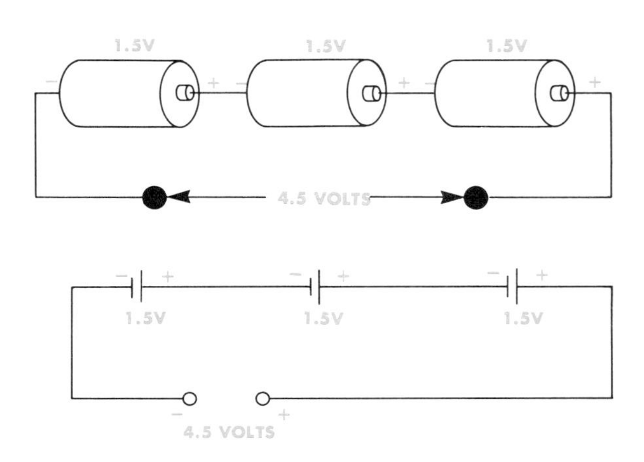

Figure 14-2: 3 batteries in series give 3 times the output voltage and the amp-hour capacity of one.

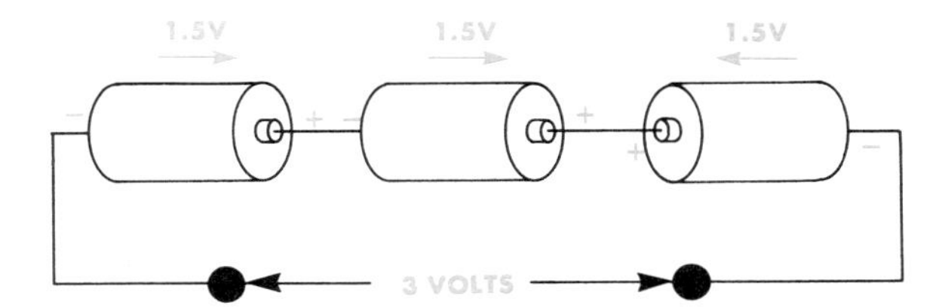

Figure 14-3: When batteries oppose each other, it doesn't hurt anything, but 1.5 volts gets cancelled out. The output voltage will be 3 volts.

Figures 14-3 and 14-4 show what happens when one or more batteries is connected in the opposite direction.

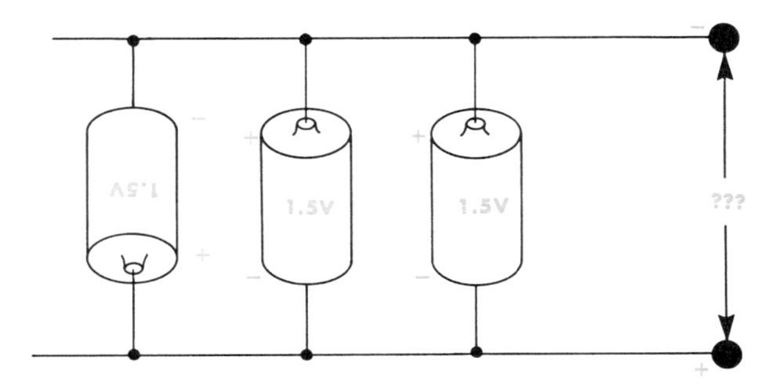

Figure 14-4: Don't do this. It is asking for problems. With real batteries you can't always be sure what will happen.

SELF CHECK

1. How many 1.5 volt cells are in a 6 volt battery? How many 2 volt cells?
2. What is the minimum and maximum voltage of a single cell?
3. How long will a 10 amp-hour battery deliver 5 amps?
4. Which will increase voltage: Wiring batteries in series or parallel?

Unit 15 SERIES AND PARALLEL CIRCUITS

All circuits must have a voltage source, a conductor for electrons to flow through, and a load. The load may be a light bulb, a transistor, or any other electrical device. Whatever the load is it has some resistance.

A resistor symbol is easy to draw but toasters and motors are not. For that reason the resistor symbol is used in drawings to mean any kind of load.

There are only two kinds of circuits:

- Series Circuits
- Parallel Circuits

A series circuit is one in which there is only one path for the electrons to flow through. If one element burns out, everything in the circuit goes out (figure 15-1).

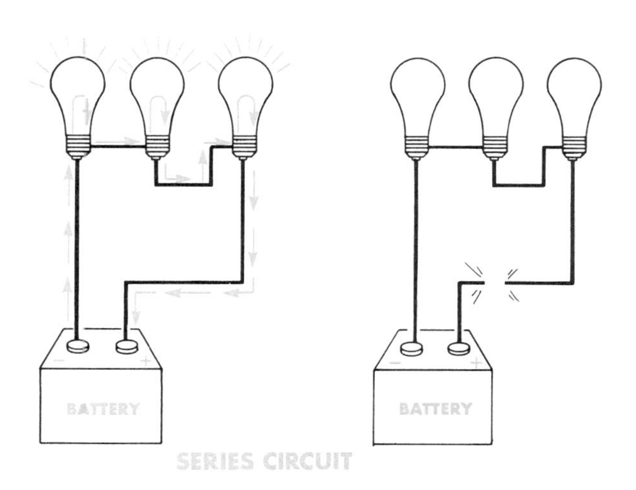

Figure 15-1: A series circuit has only one path for electrons to flow through. If it breaks or a part burns out, everything in the circuit goes out.

Parallel circuits are circuits that have more than one path for electrons to flow through. They are really several completely separate circuits that get their power from the same source. One device, or element, can burn out or be disconnected without affecting the other devices in the circuit (figure 15-2).

Figure 15-2: Household appliances are connected in parallel. Unplugging the drill will not affect the lamp. Turning the lamp on or off will not affect the drill.

When a wire breaks, or a part burns out, a gap through which electrons cannot flow is created. This is called an open circuit. Switches are used to open and close circuits conveniently (figure 15-3). When something electrical doesn't work most people say there is a short circuit (figure 15-4). However, an open circuit such as a loose connection or a broken wire is more often the trouble.

When we do have a short circuit it means that two main wires from a battery or power line are touching each other (figure 15-5). There is almost no resistance in a short circuit. This demands all the current the battery or generator can deliver. Most batteries and generators can deliver enough current to melt steel and cause fires. So short circuits can be dangerous.

Figure 15-4: Whenever something electrical doesn't work, most people think the problem is a short circuit. They are usually wrong.

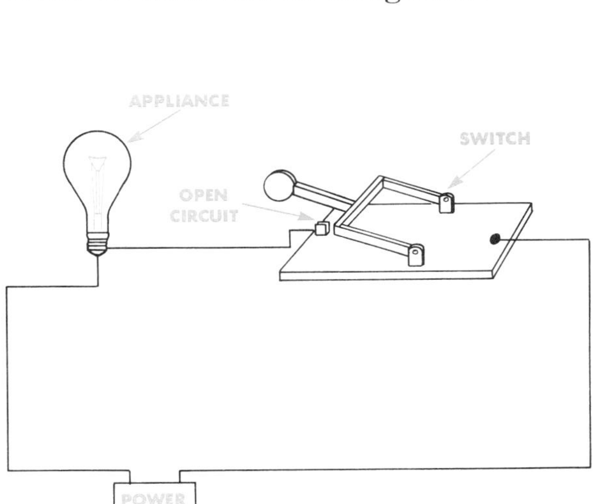

Figure 15-3: If the switch is open (off) there is no path for the electrons to flow through and the device will not work. A switch is just a convenient way to break a wire in two and reconnect it whenever we wish.

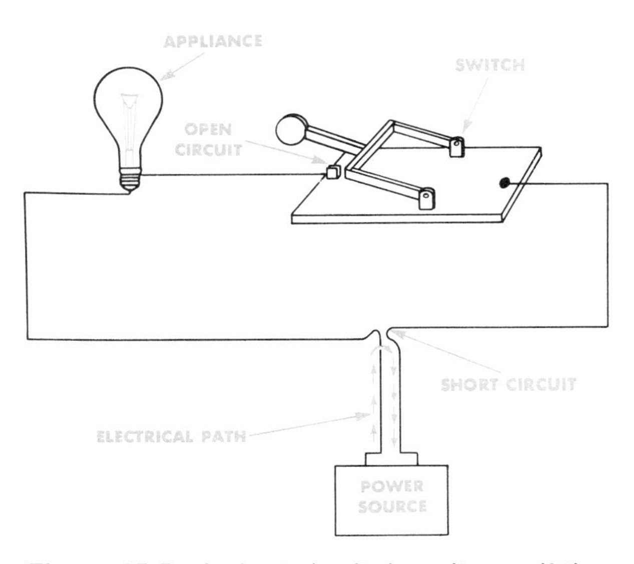

Figure 15-5: A short circuit doesn't care if the wires catch fire, the generator burns out, the battery goes dead, or your house burns down.

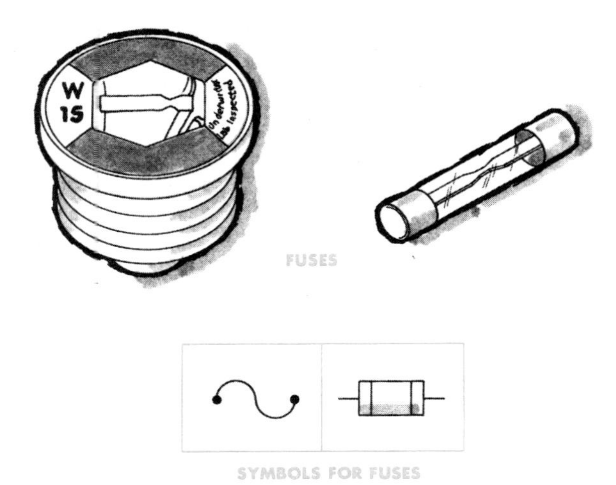

Figure 15-6: Every electrical device has resistance. As electrons move through an electrical circuit they rub against each other and create heat. When too many electrons are moving through a circuit they will create too much heat. A fuse is made so it will burn out before other circuit parts are damaged.

Fuses (figure 15-6) are safety devices that are made so they will burn out before anything is damaged by a short circuit. They are connected in series so all current flow is stopped when they burn out. A circuit breaker does the same thing by opening a special switch when it gets too hot.

SELF CHECK

1. What is a short circuit?
2. What is an open circuit?
3. If there are three light bulbs in series, what happens when one burns out?
4. What is the purpose of a fuse?

MORE ABOUT SERIES AND PARALLEL CIRCUITS — Unit 16

When electrical devices are connected in series, they must divide the supply voltage. The Christmas tree lights in figure 16-1 are connected in series to a 120 volt supply voltage. If all the bulbs have the same resistance they will divide the 120 volts equally. There are 8 bulbs, so each one will receive 1/8th of the 120 volts. That means 120 (volts) ÷ 8 (bulbs) = 15 (volts for each bulb).

Imagine that it is Christmas day and the stores are closed. Two bulbs burn out and you can't get more. Because the string is a series circuit none of the bulbs will light.

What will happen if you cut the wires, take the two sockets out, and connect the wire back together with a wire nut? You will have a series circuit with only 6 bulbs (figure 16-2), but you still have 120 volts. Now the 120 volts must be divided equally among only 6 bulbs. That means 120 (volts) ÷ 6 (bulbs) = 20 (volts for each bulb). The six bulb string will probably work for a while, but the life of the 6 bulbs will be much shorter than normal. Cut the string of lights down to 4 bulbs (120 ÷ 4 = 30), and at least one of them will probably burn out as soon as you plug it in.

You could make your series string of Christmas tree lights last longer if you replaced the missing bulbs with resistors (figure 16-3). The resistors won't give light, but they will ease the strain on the remaining 6 bulbs.

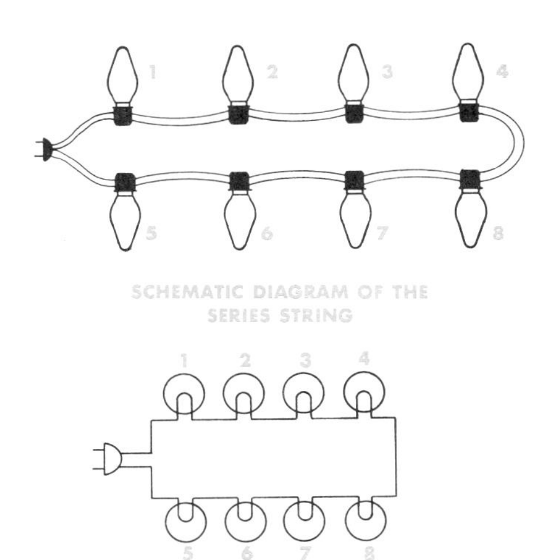

Figure 16-1: Christmas tree lights are a good example of a simple series circuit. Each bulb gets 15 volts.

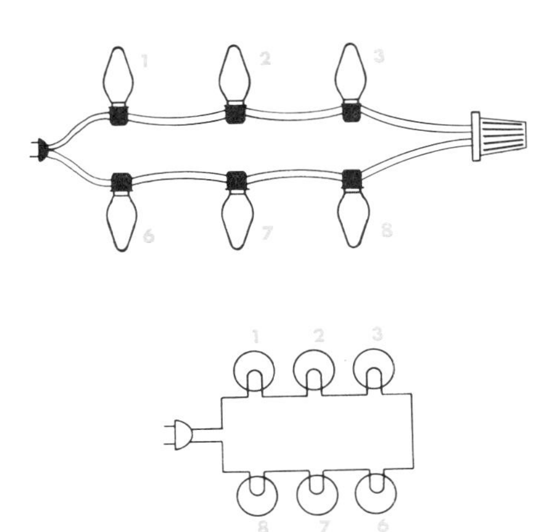

Figure 16-2: If we take the 120 volts and divide it among only six bulbs each bulb gets 20 volts.

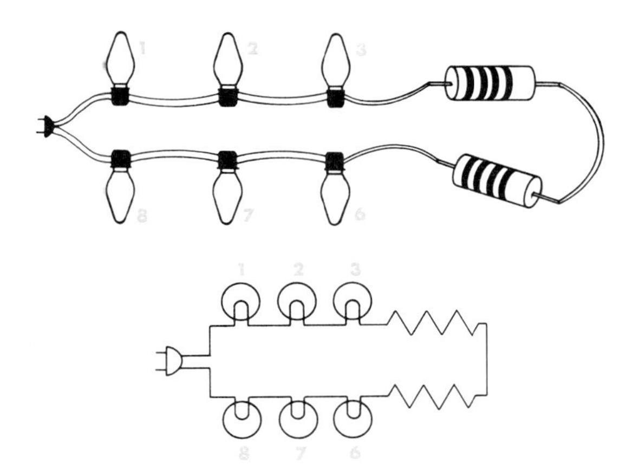

Figure 16-3: If we replace the two missing bulbs with resistors, each resistor and each bulb will each get 15 volts. You're right back where you started.

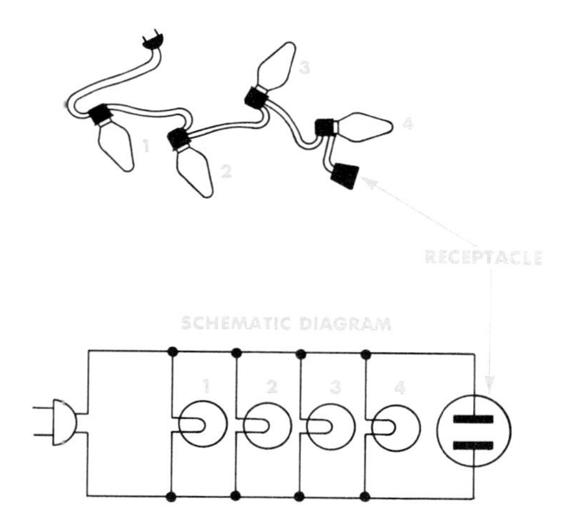

Figure 16-4: In a parallel string of Christmas tree lights each bulb has its own separate circuit. Each bulb gets the full 120 volts, so each bulb must be rated for 120 volts. If one bulb burns out it doesn't affect the others. The receptacle at the end of the string lets us add on more parallel strings.

Suppose each of the burned out bulbs has a resistance of 60 ohms. If you put two 60 ohm resistors in the circuit in place of the two burned out bulbs you would be right back where you started. The total resistance in the circuit would be equal to that of the original eight light bulbs. Each bulb would use 15 volts, each resistor would use 15 volts, and the 6 remaining bulbs would have a normal life span.

What if you replaced the two 60 ohm resistors with one 120 ohm resistor?

When you connect resistors in series you add their values. Two 60 ohm resistors in series have the same resistance as one 120 ohm resistor ($60 + 60 = 120$). So, a 120 ohm resistor can be used to replace two 60 ohm resistors in a series circuit. The voltage across it will be 30 volts ($2 \times 15 = 30$).

SELF CHECK

1. What is the voltage of each light on a 12 light string plugged into 120 volts?
2. What happens when a bulb burns out in a series string?
3. What happens in a parallel string if one bulb burns out?
4. Can a single resistor of 100 ohms replace 2 bulbs of 50 ohms each?

RESISTORS IN SERIES AND PARALLEL

The method for finding resistor values varies by the way they are wired into a circuit. They can be wired in:

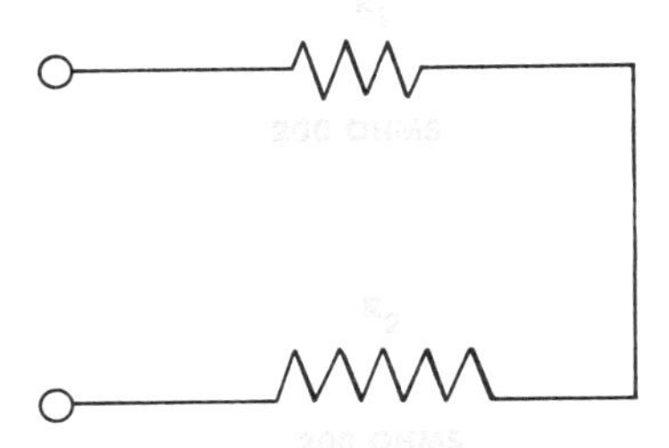

Figure 17-1: To find the total value of the resistors in a series circuit we simply add their individual values. In this case 200 ohms plus 200 ohms equals 400 ohms.

- Series
- Parallel
- Series—Parallel

As we learned in Unit 16, to find the total value of resistors connected in series, we simply add their individual values (figure 17-1). In a series circuit with one 150 ohm resistor, and one 100 ohm resistor, the total resistance of the two resistors is 250 ohms (150 ohm + 100 ohms = 250 ohms). Both resistors could be replaced by one 250 ohm resistor. If three resistors are wired in series, and their total resistance adds up to 425 ohms, they can be replaced by a single 425 ohm resistor (figure 17-2).

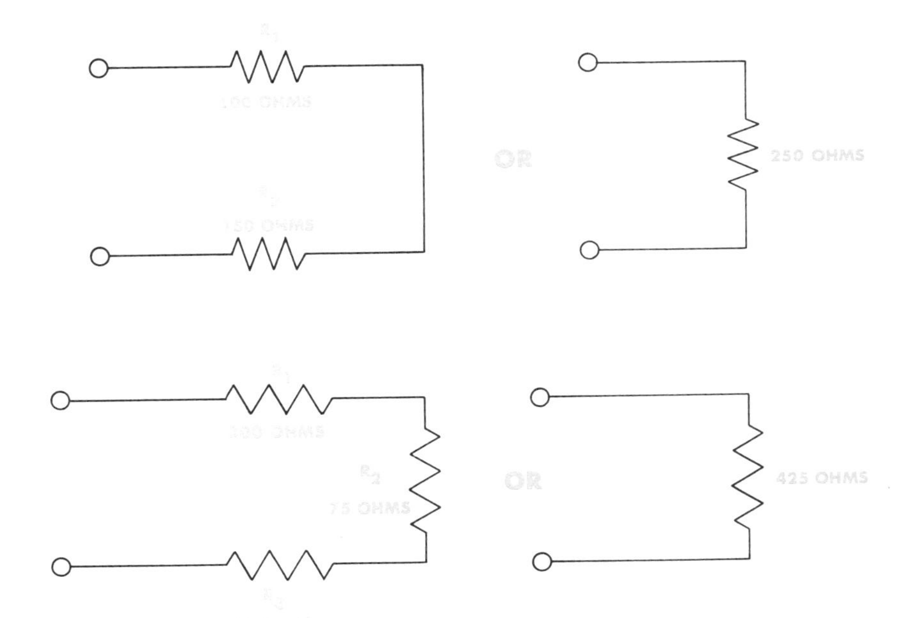

Figure 17-2: In a series circuit, a single resistor can be used to replace two (or more) resistors.

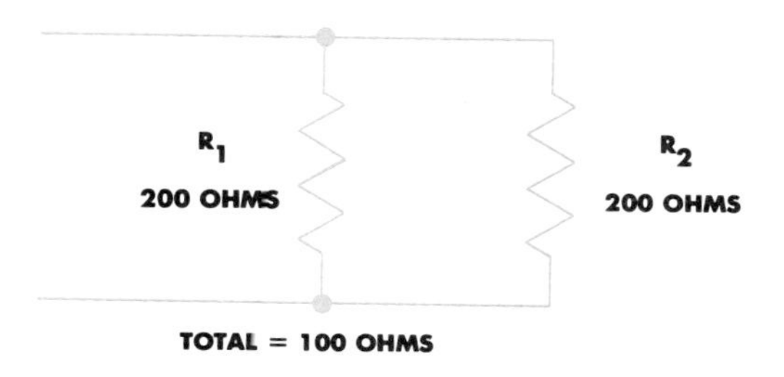

Figure 17-3: To find the total value of the resistors in a parallel circuit when all values are the same we simply divide the value of one resistor by the number of resistors in the circuit. In this case 200 ohms divided by 2 resistors equals 100 ohms.

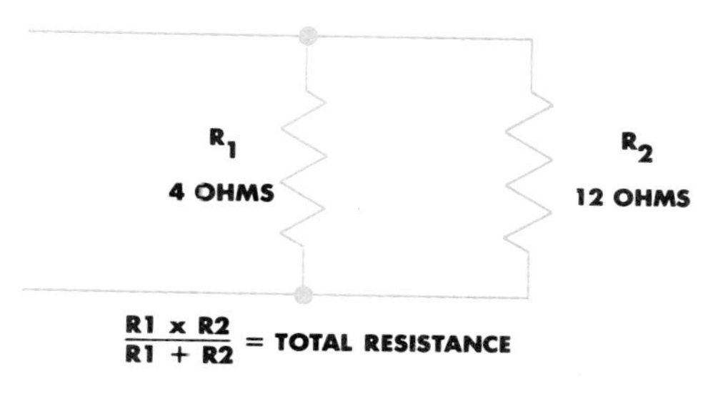

Figure 17-4: How to find the total value of the resistors in a parallel circuit when their values are not the same.

$$\frac{R_1 \times R_2}{R_1 + R_2} = \frac{4 \times 12}{4 + 12} = \frac{48}{16} = 3 \text{ ohms}$$

Resistors in parallel circuits are a bit more complicated.

If all the resistors in a parallel circuit have the same resistance value, just divide the value of one by the total number of resistors in the circuit (figure 17-3). If they have different values we must use the formula shown in figure 17-4.

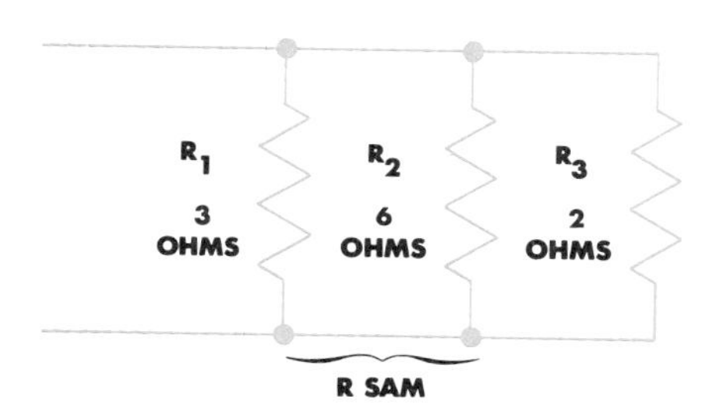

Figure 17-5: Here are the two steps it takes to solve this problem:

1. $\frac{R_1 \times R_2}{R_1 + R_2} = \frac{3 \text{ ohms} \times 6 \text{ ohms}}{3 \text{ ohms} + 6 \text{ ohms}} = \frac{18 \text{ ohms}}{9 \text{ ohms}} = 2 \text{ ohms}$

   2 ohms $= R_{SAM}$

2. $\frac{R_{SAM} \times R_3}{R_{SAM} + R_3} = \frac{2 \text{ ohms} \times 2 \text{ ohms}}{2 \text{ ohms} + 2 \text{ ohms}} = \frac{4}{4} = 1$

The total resistance in this circuit is 1 ohm.

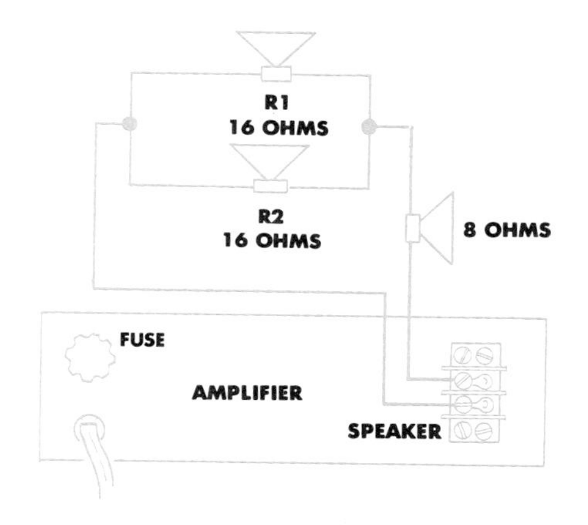

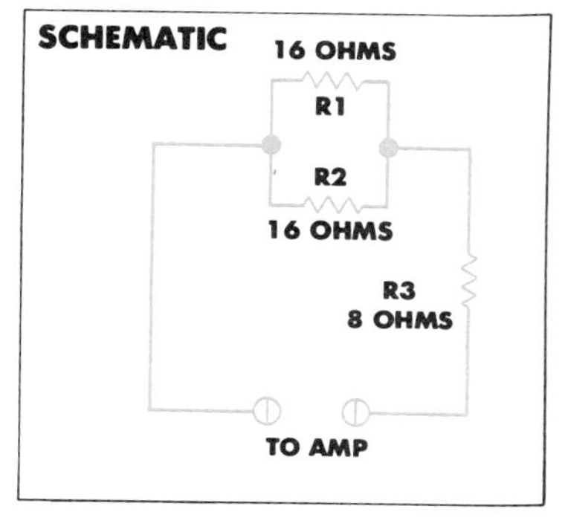

Figure 17-6: Here's a problem you might run into when you hook up your stereo system. Can these three speakers replace one 16 ohm speaker?

This formula is only good for two resistors each time we use it. If there are more than two resistors in the parallel circuit we're working on we must use the same formula over and over until we work it for each resistor (figure 17-5). Begin by finding the total resistance of the first two resistors (R_1 and R_2):

$$\frac{R_1 \times R_2}{R_1 + R_2} = \frac{3 \text{ ohms} \times 6 \text{ ohms}}{3 \text{ ohms} + 6 \text{ ohms}} = \frac{18 \text{ ohms}}{9 \text{ ohms}}$$

$$\frac{18 \text{ ohms}}{9 \text{ ohms}} = 2 \text{ ohms}$$
(total resistance of R_1 and R_2)

Now we have to work the problem over again to find out how R_3 affects the circuit. To make it a little easier, let's call the answer to our first equation RSAM. All we have to do is exchange RSAM for R_1, and R_3 for R_2 in the original formula, and work it over again. So:

$\frac{R_1 \times R_2}{R_1 + R_2}$ is changed to $\frac{R_{SAM} \times R_3}{R_{SAM} + R_3}$

$$\frac{R_{SAM} \times R_3}{R_{SAM} + R_3} = \frac{2 \text{ ohms} \times 2 \text{ ohms}}{2 \text{ ohms} + 2 \text{ ohms}} = \frac{4 \text{ ohms}}{4 \text{ ohms}}$$

$$\frac{4 \text{ ohms}}{4 \text{ ohms}} = 1 \text{ ohm}$$

1 ohm is the total resistance of these three resistors when they are wired in parallel.

Some circuits are part series and part parallel. This kind of circuit is called a series—parallel circuit (figure 16-6). There's nothing really complicated about them. All we have to do is find the total resistance of each parallel part of the circuit with the R_1 and R_2 formula. When we've found the single value for each parallel part of the circuit, we can think of it as a single resistor. What we have left is a simple series circuit. All we need to do is add up the values and we have the answer.

SELF CHECK

1. What is the total resistance of this circuit?

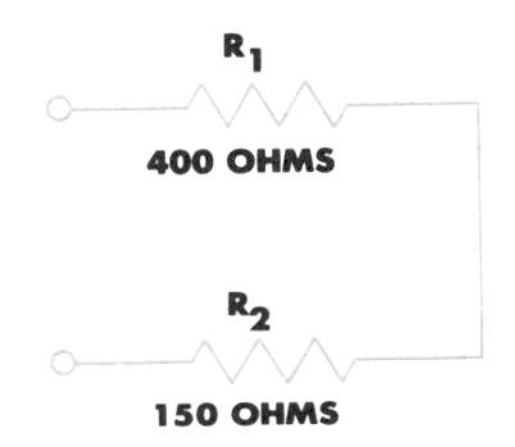

2. What is the total resistance of this circuit?

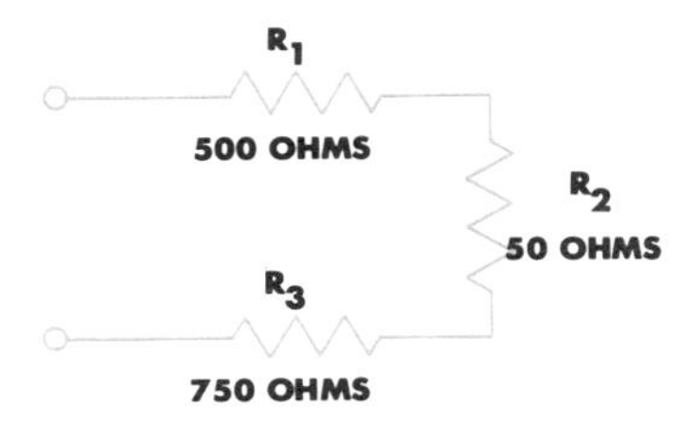

3. What is the total resistance of this circuit?

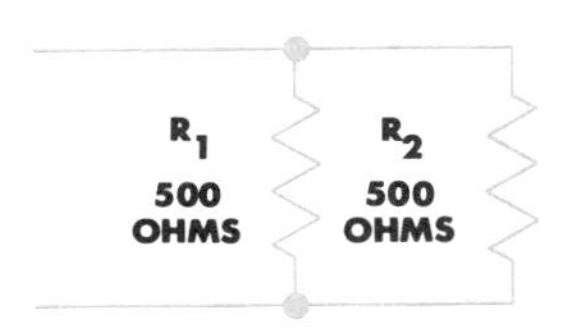

4. What is the total resistance of this circuit?

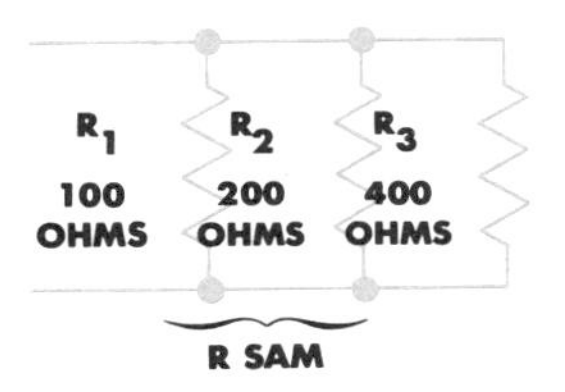

Unit 18 MAGNETS

The electrons spinning around the nucleus of <u>all</u> atoms have small magnetic fields. Just like the earth, each of these tiny magnetic fields has its own north pole and its own south pole (figure 18-1).

The electrons in most non-magnetic materials spin and rotate randomly, so the tiny magnetic fields of their electrons cancel each other out. In magnetic materials, like iron, the atoms join together in groups called <u>domains</u> (figure 18-2). A domain is a tiny magnet made up of a lot of electron magnets. The electrons in a domain spin together, and each adds its magnetic strength to the group.

Domains can be lined up so that they all point in the same direction. When all the domains in a piece of iron are lined up the piece of iron becomes a magnet (figure 18-3).

If we break a large magnet into small pieces we will have several smaller magnets because the domains will still be lined up (figure 18-4).

Magnetic fields always affect one another in the same ways:

- Opposite fields attract each other
- Like fields repel each other

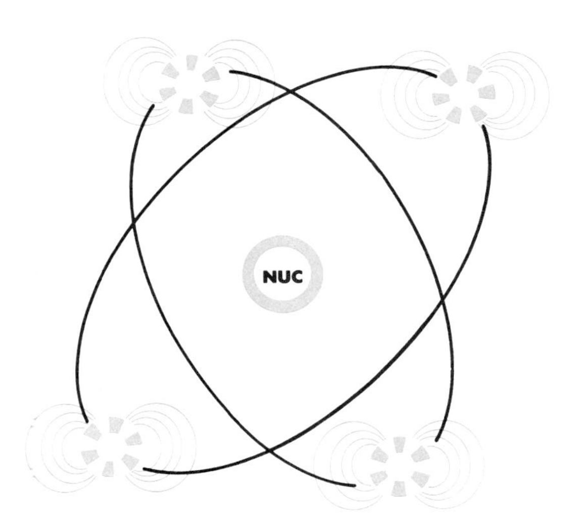

Figure 18-1: Each electron, as it spins around the nucleus, acts like a tiny magnet.

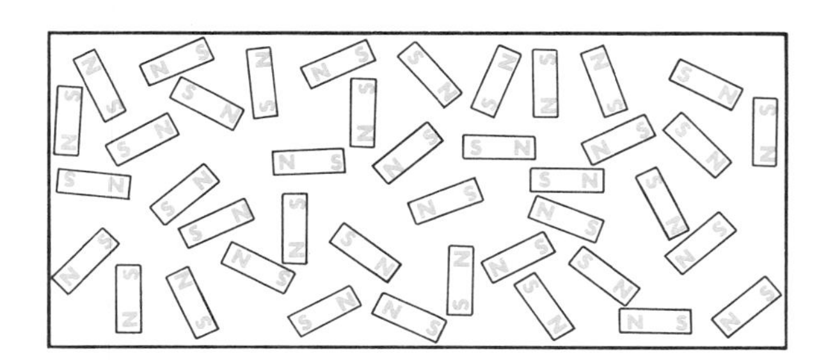

Figure 18-2: In iron, atoms join to form mini-magnets called domains.

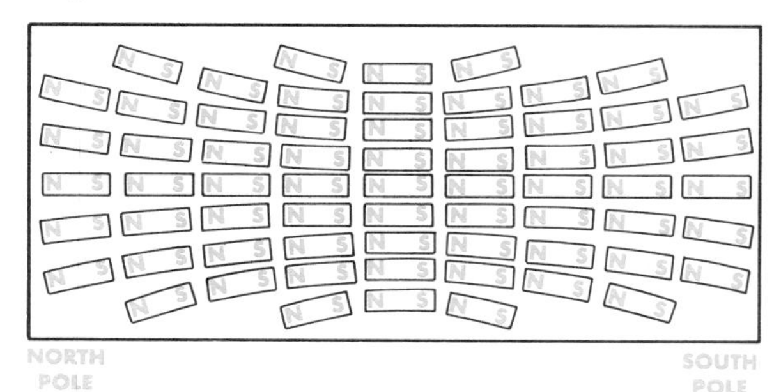

Figure 18-3: When the domains are lined up, the piece of iron becomes a magnet.

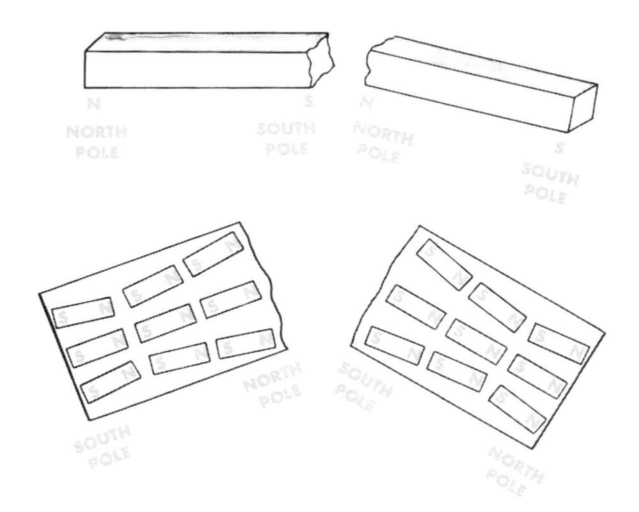

Figure 18-4: When we break a magnet, we get two magnets. Each one has its own North and South pole.

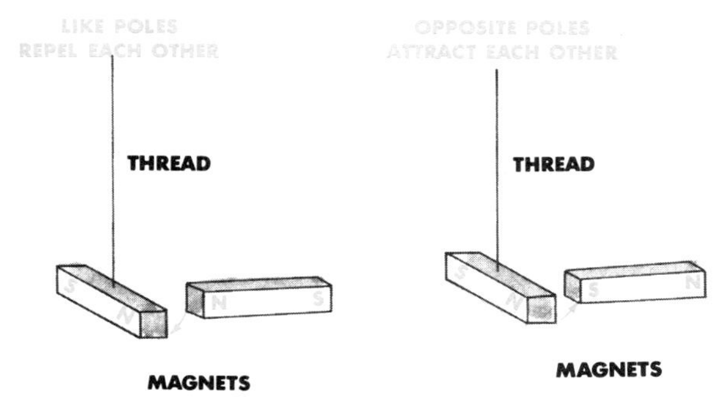

Figure 18-5: Opposite poles attract. Like poles repel.

This means that the fields surrounding the north poles of two magnets will push against each other, as will those surrounding the two south poles. The north pole of one magnet and the south pole of another will attract each other (figure 18-5).

When a magnet gets close to a piece of iron or steel, the domains in the iron or steel line up with the domains in the magnet, and the piece of iron or steel becomes a magnet. Since opposite fields attract each other, the iron or steel sticks to the magnet. It works the same with any magnetic material.

When the magnet is taken away, the domains in iron return to their original state very quickly, but the domains in steel stay lined up. The steel then becomes a permanent magnet.

SELF CHECK

1. What causes the smallest magnetic field?
2. What are domains made of?
3. How does breaking a magnet affect its magnetism?
4. What does "likes repel, opposites attract" mean?

Unit 19 ELECTROMAGNETS

Electrons orbiting an atom produce tiny magnetic fields. If we push electrons through a wire there will also be a field around the wire (figure 19-1).

It doesn't matter what makes the electrons move. When they are moving they produce a magnetic field.

If we wind a long wire into a coil we can get a large number of electrons working together to produce a stronger magnetic field (figure 19-2).

If we put a piece of iron into the core of the coil, the iron domains will line up. There are now two magnets working together, the coil and the iron

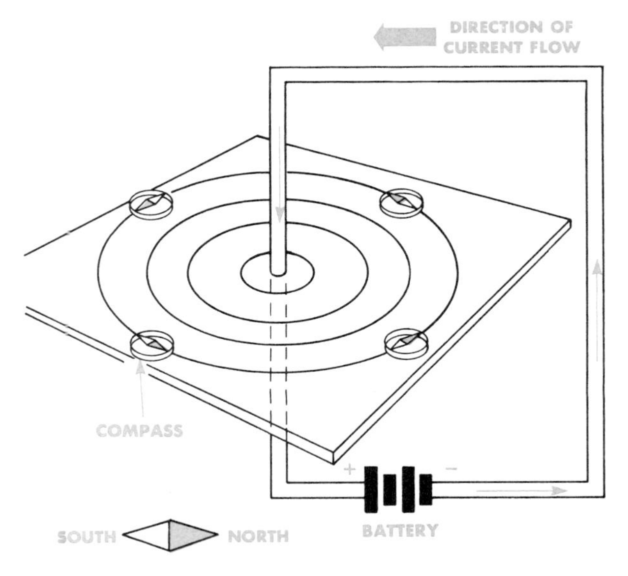

Figure 19-1: Electrical current flowing through a wire sets up a magnetic field. A compass will swing to show the direction of the field.

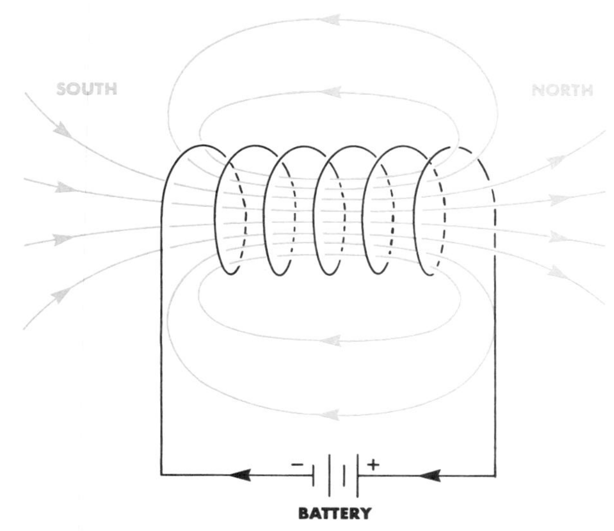

Figure 19-2: If we wind a long piece of wire into a coil we will get an electromagnet strong enough to be useful.

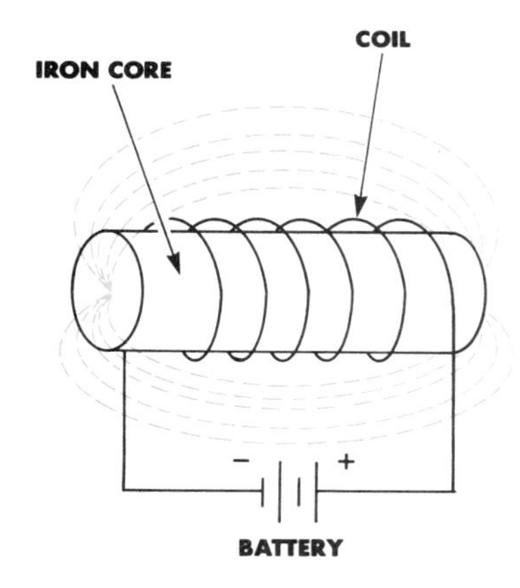

Figure 19-3: A piece of iron in the center of the coil can make the electromagnet 1000 times stronger.

magnet in its core. We can make powerful electromagnets in this way (figure 19-3).

Two other ways to increase the strength of an electromagnet are:

- More turns of wire on the coil makes a stronger electromagnet
- More current also increases the strength of the electromagnet

When we turn off the electrical current through the coil, the magnetic field around the coil vanishes. The iron domains no longer have the coil's field to keep them lined up. They very quickly fall out of line. The electromagnet will no longer pick up iron.

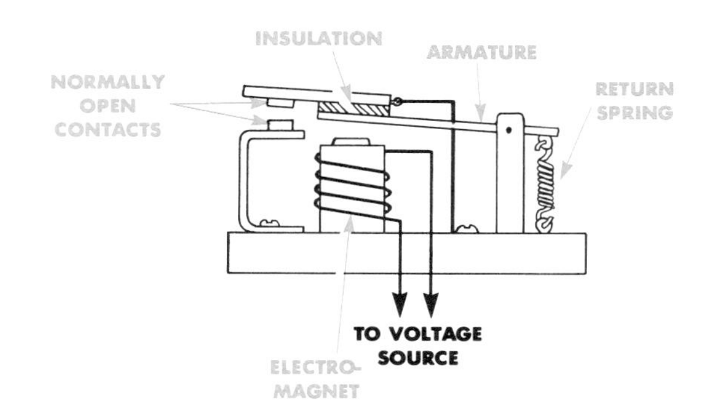

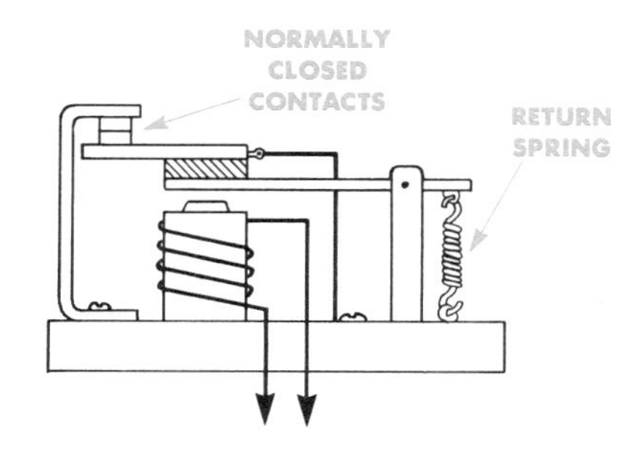

Figure 19-4: A small amount of power applied to the coil of a relay can turn high power motors, lights, etc.,on or off.

The relay (figure 19-4) is an important application of the electromagnet. The relay got its name when it was used in early telegraph systems.

When the signal had traveled far enough on the telegraph wire it got so weak that it couldn't work the telegraph receiver.

The relay could work with the weak signal and switch in another battery. The signal could then start out strong for the next 25 mile run.

This is something like passing the baton in a relay race to a fresh runner. In the telegraph the signal was passed to a fresh battery.

The relay has a coil with an iron core. When a current flows through the coil it pulls an iron bar down to it. Switch contacts are attached to the bar. When current stops flowing through the coil, a spring pulls the bar back up. This opens the switch.

Relays are used in many modern applications. The relay can be used wherever a little bit of power must turn a large power on or off.

SELF CHECK

1. True or False: Electrons moving through a wire set up a magnetic field.
2. Is the strength of an electromagnet increased when an iron core is used?
3. List three ways to make a stronger electromagnet.
4. Why is a long piece of wire wound into a coil?

Like poles repel and opposite poles attract. Electric motors (figure 21-1) are machines that use these two simple principles to push and pull an electromagnet mounted on a rotating shaft. This electromagnet is called an armature. Two more magnets (called the field) are mounted on the motor frame of an electric motor. The field magnets must be mounted opposite each other.

Special sliding switch contacts (called brushes) are needed to connect the voltage source to the rotating coil of the electromagnet. Wires can't be hooked directly to the rotating coil because the spinning shaft would twist them until they broke. Figure 20-2 shows a simple set-up that is used in some motors.

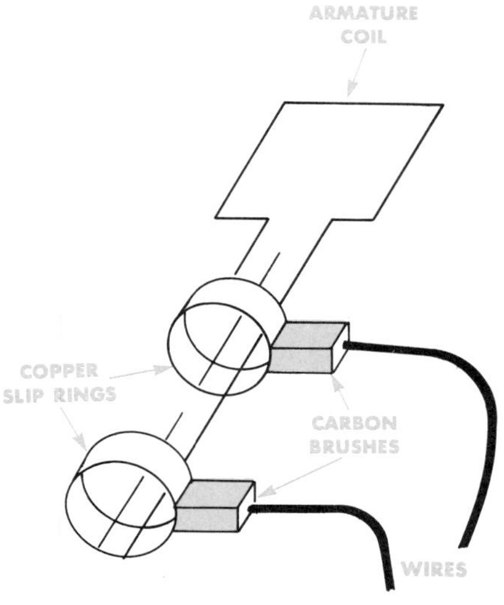

Figure 20-2: Brushes are used to make electrical connection to a rotating coil.

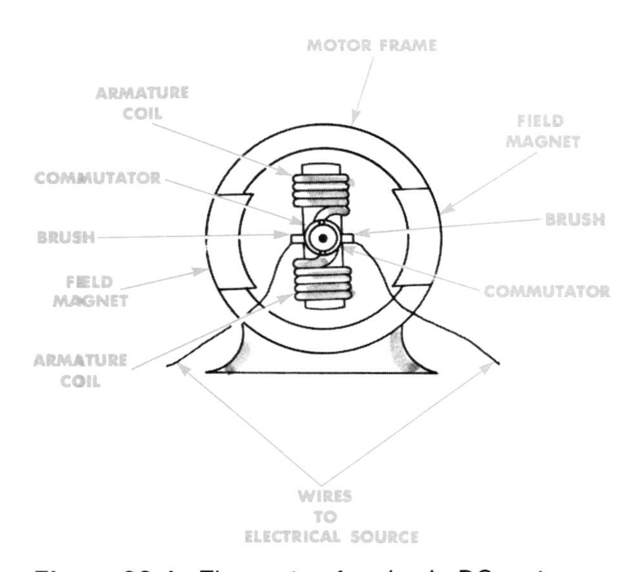

Figure 20-1: The parts of a simple DC motor.

To make electrical contact between the electric source and the rotating armature electric motors must have:

- Brushes
- Slip-rings (for many AC motors)
- Commutator (for most DC motors and many AC)

Slip-rings and commutators are copper rings that are mounted on the rotating shaft. Commutators are used on motors where the current must be

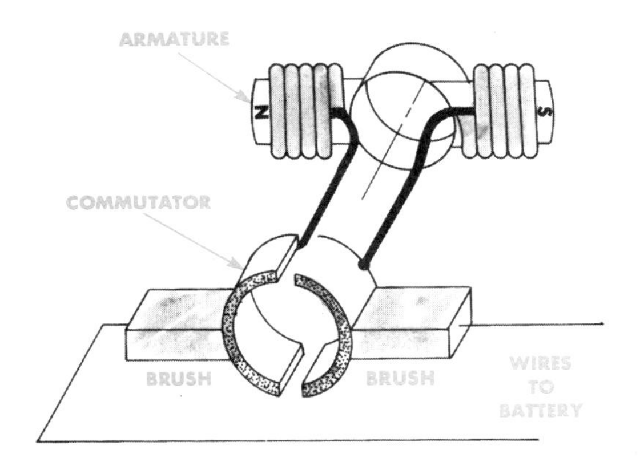

Figure 20-3: The split ring commutator is a combination of slip rings and a rotating switch.

reversed (switched) as the motor turns. Slip-rings are used on motors where switching is not required. The brushes slide on these rings. Brushes are made of graphite, a form of carbon. Carbon is not the best conductor we can use for brushes, but it provides its own lubrication. Oil or grease would act as an insulator and interfere with the electrical contact.

To keep a motor rotating, the positions of the battery's positive and negative wires must be switched from one

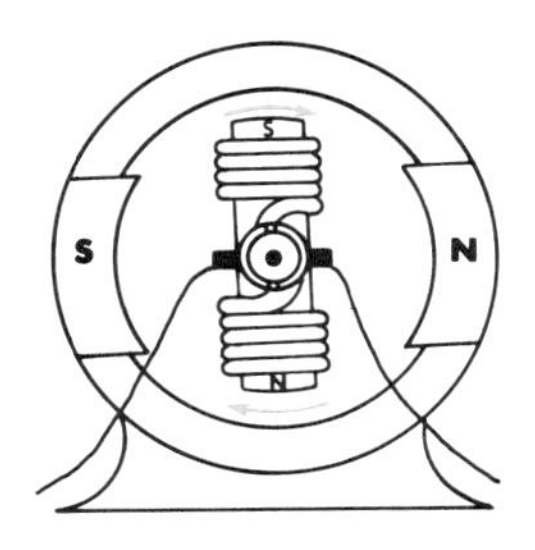

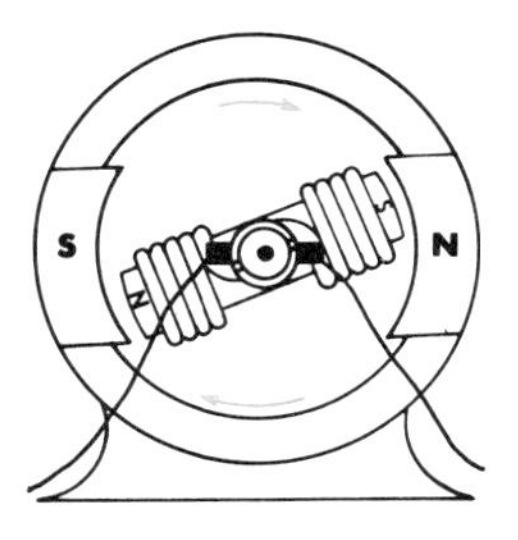

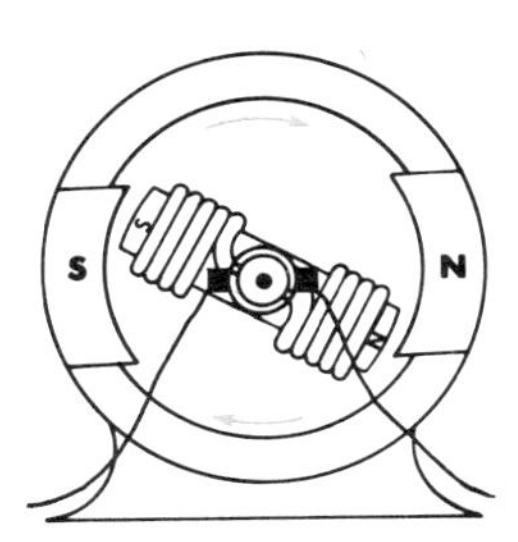

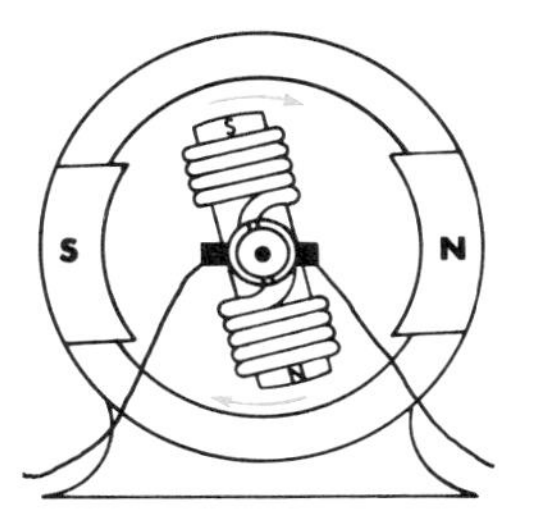

Figure 20-4: This is how an electric motor works.

side of the armature to the other during each revolution. The commutator is used to do this (figure 20-3). It does it by acting as a rotary switch. As the shaft rotates, the commutator reverses the battery wires to change the polarity of the armature and keep the motor turning (figure 20-4).

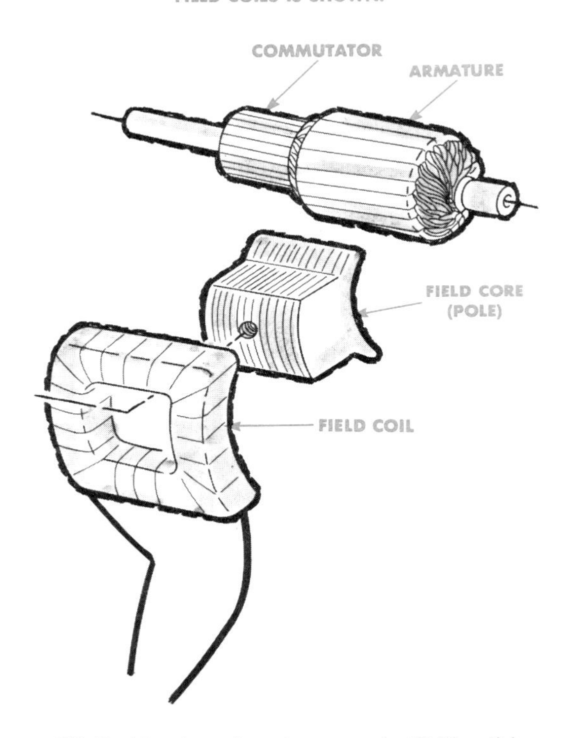

Figure 20-5: Most real motors are built like this.

Most motors have several armature coils. These extra armature coils are needed to smooth out and increase the power of the motor as it turns. Many very powerful motors also have several field coils. This requires a more complicated armature than the simple motors we have explained here but the principles are the same. It also requires a commutator with more segments.

Many motors use electromagnets for the field instead of permanent magnets (figure 20-5). Just as they increase the power of a magnet, these electromagnetic coils also increase the power of a motor.

The armature is the same for both permanent magnet and electromagnetic field motors.

SELF CHECK

1. What are the magnetic laws that make a motor work?
2. Why is a commutator needed?
3. Why are graphite (carbon) brushes used for motors?
4. Do all motors use electromagnets for their fields?

Figure 21-1 shows how a typical generator is made. What? You say there must be some mistake? The generator looks exactly like the motor in the last unit. You are right, but there is no mistake.

Commutator type motors and generators are all built the same. Almost any motor with a commutator works just as well as a generator. And, a generator with a commutator can be used as a motor.

In a motor, an electric current produces a magnetic field that moves the armature.

This also works backwards. The moving armature can push electrons through a wire to make an electric current.

To generate a current we must find some way to turn the shaft. The amount of mechanical energy we put in must be a little greater than the electrical energy we get out. Some of the mechanical energy is lost in heat because of friction.

If you ride a bicycle with a generator to run the light you can feel the extra energy required. When you turn on the light you will have to start pumping harder.

A generator works because a current will flow when we move a magnet back and forth in a coil. Current will flow through the coil and any electrical device connected to it.

In the drawing in figure 21-2, current will flow through the lamp as long as

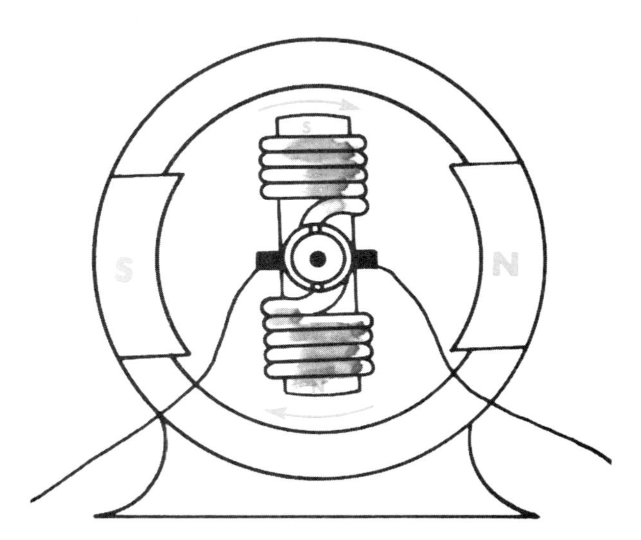

Figure 21-1: How a generator is made.

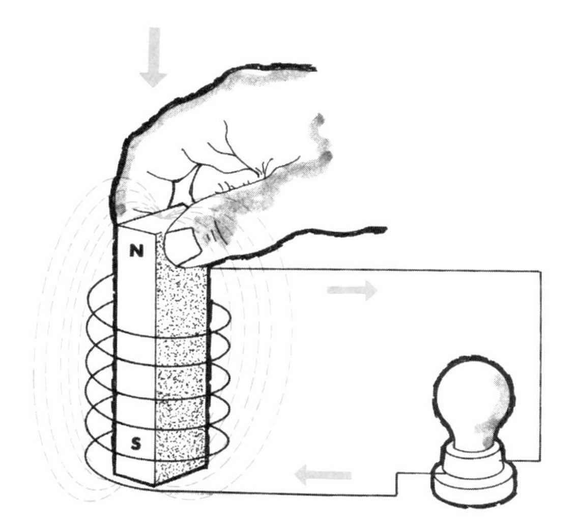

Figure 21-2: Moving a magnet in and out of a coil generates an electrical current.

we keep the magnet moving. The same thing happens if we hold the magnet steady and move the coil. The faster we move the magnet or coil the brighter the lamp gets. When we stop the movement the lamp goes out. Practical generators rotate instead of moving back and forth. The principle is the same. You will learn about alternating current generators in the next unit.

SELF CHECK

1. Can a motor often be used as a generator?
2. Will a generator produce an electric current when the shaft is not rotating?
3. Does a fast turning generator produce a higher output voltage than one that truns slowly?
4. True or false: The rotating part of a generator is called the armature.

The alternating current generator is built exactly like the direct current generator except it uses slip rings instead of a split ring commutator (figure 22-1).

The drawings in figure 22-2 show a graph of generator output voltage for each armature position. The output voltage is lower when the armature and field magnets are far apart and higher when they are close together.

A graph with the shape shown in figure 22-3 is called a sine wave. The generator is an alternating current generator. The graph describes the behavior of almost all alternating currents.

Alternating current flows through a wire or circuit in one direction and then the other.

Ordinary household electricity reverses 60 times a second. This means

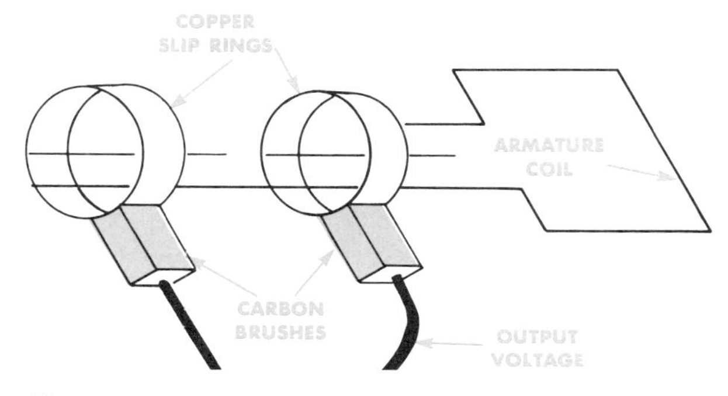

Figure 22-1: The alternating current generator is built exactly like the direct current generator except that it uses slip rings instead of a split ring commutator.

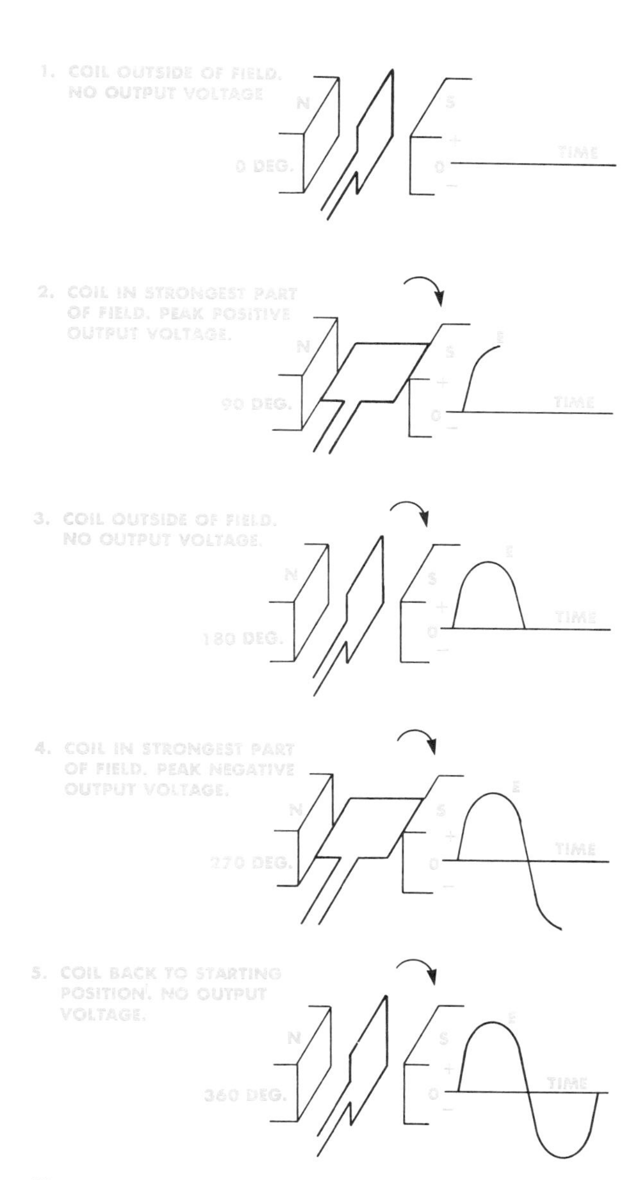

Figure 22-2: Because the output voltage is higher when the armature is closer to the magnet the output voltage constantly changes as the armature rotates.

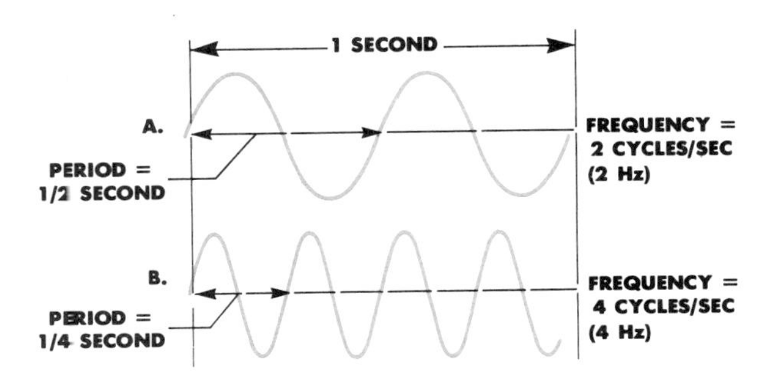

Figure 22-3: To get RMS volts, multiply Peak volts by .7 (top). To get Peak volts, multiply RMS volts by 1.4 (bottom).

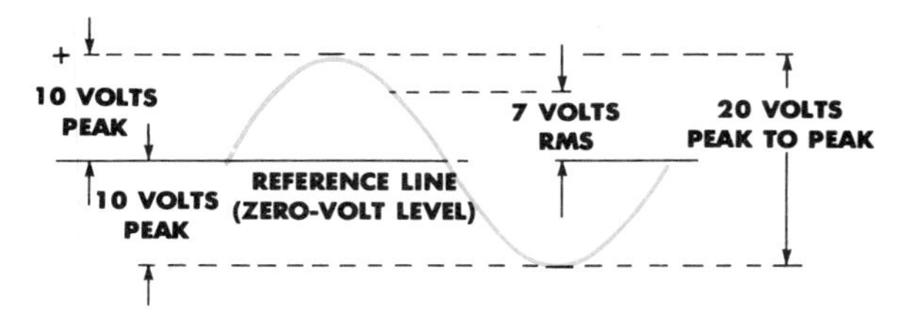

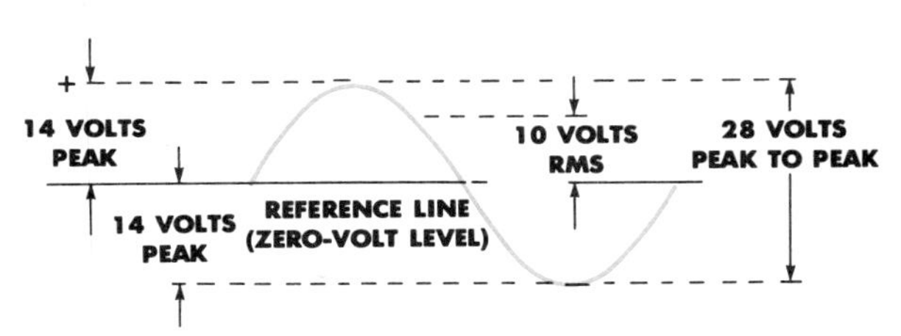

Figure 22-4: The frequency and period of a sine wave.

the generator makes 60 complete revolutions each second. This is the same as 3600 revolutions per minute.

Alternating current is not steady. It can do a lot of work when the alternating voltage is high and none at all when it is down to zero.

How do we tell someone what the voltage is when it is always changing? We could use the maximum or peak voltage. Or, we could use the voltage from one peak to the other. Both of these are used sometimes. The most common way is to use an average value. The average value is 70% of the peak voltage. The average value is obtained mathematically and is called the RMS value. RMS stands for Root Mean Square which describes the mathematical process used. Everyone just calls it RMS (figure 22-4).

The frequency of the power line in your home is 60 cycles per second. The abbreviation Hz (Hertz) is now used to mean "Cycles Per Second".

The period of a sine wave is the time it takes to complete one complete cycle. The period of the AC power line in your home is 1/60 of a second.

Radio waves, television waves, and even light waves are sine waves. The frequencies of these waves are much higher than the power line frequency. The frequency of television channel 12 is 204,000,000 cycles per second (204 million Hertz).

SELF CHECK

1. What is the frequency of the AC line in your home?
2. The wave produced by an alternating current generator is called a ________ wave.
3. If the peak of an AC voltage is 168 volts, what is the RMS voltage value.
4. If the RMS voltage is 110 volts, what is the peak voltage?

The instrument you are most likely to find in any electronics shop is a multimeter. A multimeter is often called a VOM for Volt-Ohm-Milliamp. There is a non-electric version and two electronic versions of the VOM. The model that uses vacuum tubes is called a vacuum tube voltmeter (VTVM). The kind that uses transistors is called a solid-state VOM.

The scales and settings can be a little confusing at first, but a little practice will take care of the problem. Figure 23-1 shows a typical VOM.

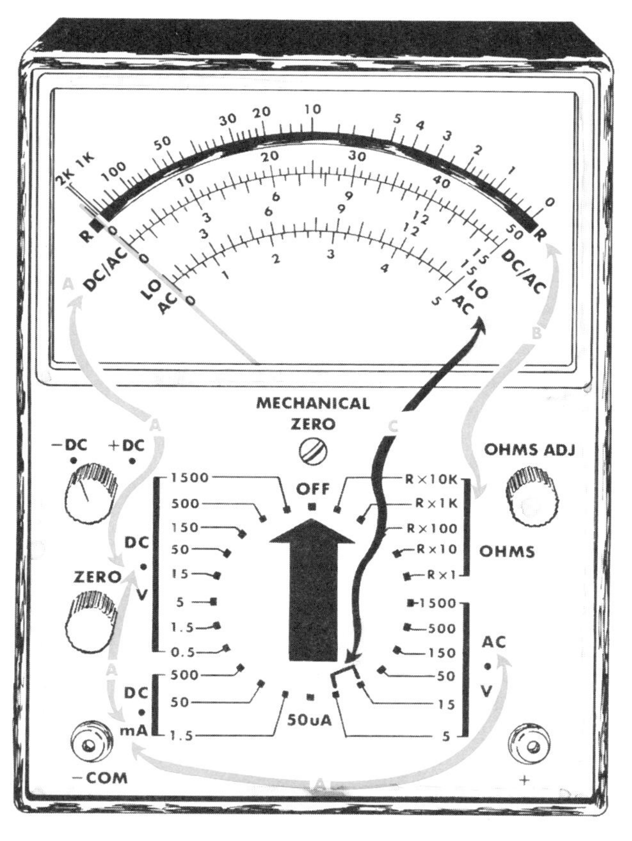

Figure 23-1: There are three scales on a typical VOM. Scale A is used to measure DC voltage, AC voltage and milliamperes. Scale B is used only for ohms. Scale C is a special scale used to measure AC voltages of less than 15 volts.

There are three scales. Scale A is used for DC volts, AC volts, and milliamperes. Scale B is used only for Ohms. Scale C is a special scale for AC measurements of less than 15 volts.

A device called a meter rectifier is used to convert AC volts into DC volts. The meter rectifier doesn't cause any problems at higher voltages, but below 15 volts it causes a large error. The problem is solved by using a special scale for low AC volts.

Reading DC voltage is easy. The pointer knob in figure 23-2 is set on 500. 500 volts is the highest voltage that can be measured on this range. Follow the arrowed line on the drawing up to the meter scale. The number there is 50, but the pointer knob reads 500. This means you must add a zero to all readings on the scale. Now look at the needle. It is half way between 200 and 300 volts. (Don't forget to add the zero). The reading is about 250 volts. The little arrows by the pointer knob show what other ranges must be read on this same scale. In each case, you must add or remove zeros or put in the decimal point.

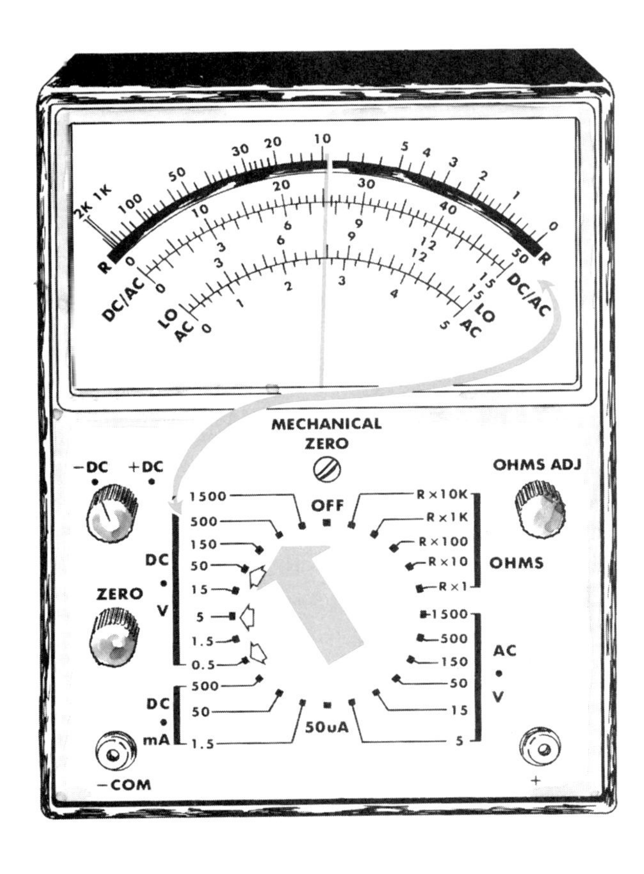

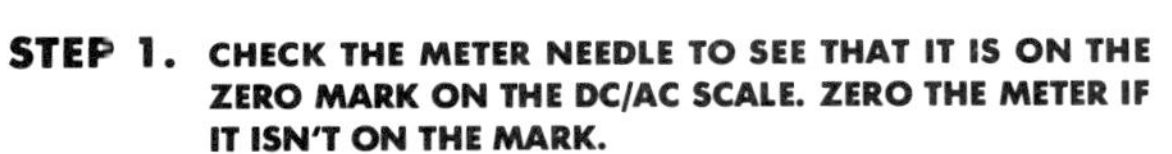

STEP 1. CHECK THE METER NEEDLE TO SEE THAT IT IS ON THE ZERO MARK ON THE DC/AC SCALE. ZERO THE METER IF IT ISN'T ON THE MARK.
STEP 2. SET THE POINTER KNOB TO 1500 DC VOLTS.
STEP 3. CONNECT THE TEST LEADS (−COM AND +) TO THE VOLTAGE TO BE MEASURED. IF THE POINTER MOVES TO THE LEFT OF ZERO (BACKWARDS) EITHER EXCHANGE THE TEST LEADS OR TURN THE −DC +DC KNOB.
STEP 4. TURN THE POINTER KNOB TO LOWER VOLTAGE RANGES UNTIL YOU GET A READING WHERE THE NEEDLE IS AT LEAST A FEW MARKS AWAY FROM EITHER EDGE OF THE SCALES.

Figure 23-2: Using a VOM to measure DC voltage.

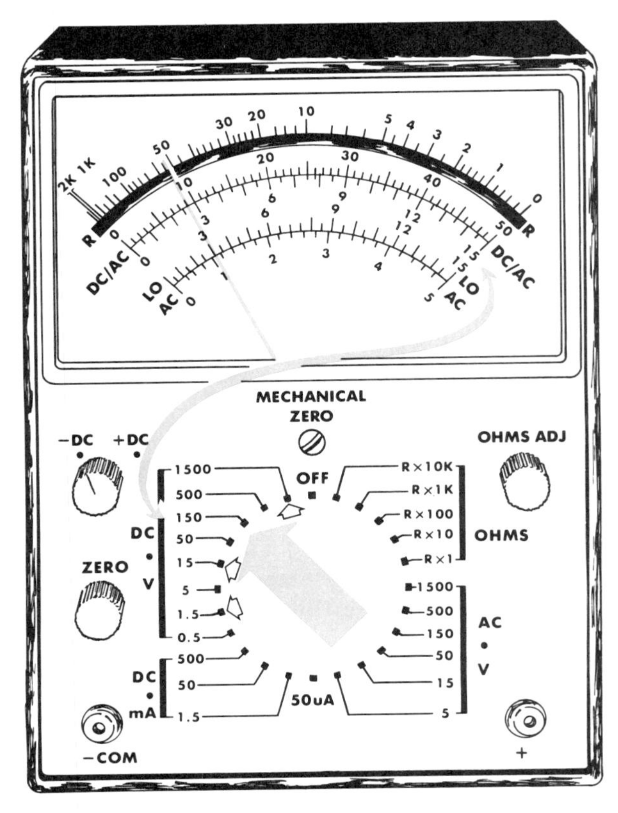

Figure 23-3: Different scales require different calculations.

Figure 23-3 is another example of how to read DC volts. The pointer knob is set on 150. 150 volts is the highest voltage that can be measured on this range. Follow the arrowed line on the drawing up to the meter scale. The number there is 15, but the pointer knob reads 150. This also means you must add a zero to all readings on the scale. Now look at the needle. It is very close to 30 volts. (Don't forget to add the zero.) The reading is about 29 volts. The little arrows by the pointer knob show what other ranges must be read on this same scale. Don't forget to add or remove the zeros or put in the decimal point.

AC volts are measured using the same method we use to measure DC volts. The only difference is that AC voltages below 15 volts must be read on the LO AC meter scale instead of the DC/AC scales.

Current (milliamps) is measured just like DC volts except that the pointer knob will be set to the DC mA ranges.

Figure 23-4 shows how to use a VOM to measure resistance.

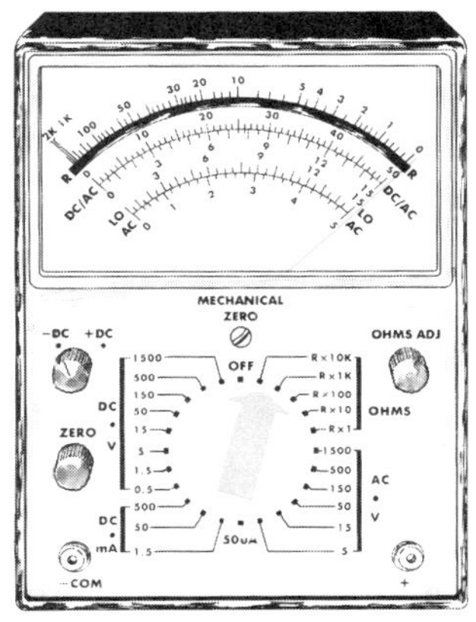

STEP 1. SET THE POINTER KNOB TO RX10K.

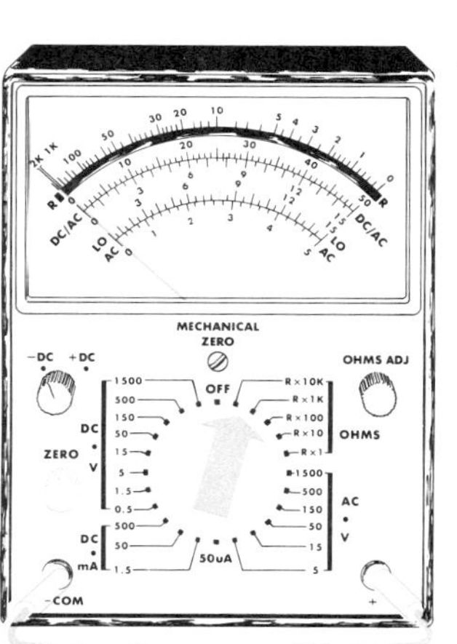

STEP 2. HOLD THE −COM AND + TEST LEADS TOGETHER.

STEP 3. THE METER NEEDLE SHOULD SWING TO THE ZERO MARK ON THE OHMS SCALE. IF THE NEEDLE IS NOT RIGHT ON THE ZERO MARK, TURN THE ZERO KNOB UNTIL IT IS. SOME METERS DO NOT HAVE A ZERO KNOB. YOU WILL HAVE TO USE THE MECHANICAL ZERO SCREW ON THAT KIND OF METER. DON'T USE THE MECHANICAL ZERO IF THE METER HAS A ZERO KNOB.

STEP 4. SEPARATE THE TEST LEADS SO THEY NO LONGER TOUCH. THE NEEDLE SHOULD SWING TO THE OPPOSITE SIDE OF SCALE. AND LINE UP WITH THE ∞ MARK.

STEP 5. IF THE POINTER DOES NOT LINE UP WITH THE ∞ MARK, TURN THE OHMS ADJ KNOB UNTIL IT DOES.

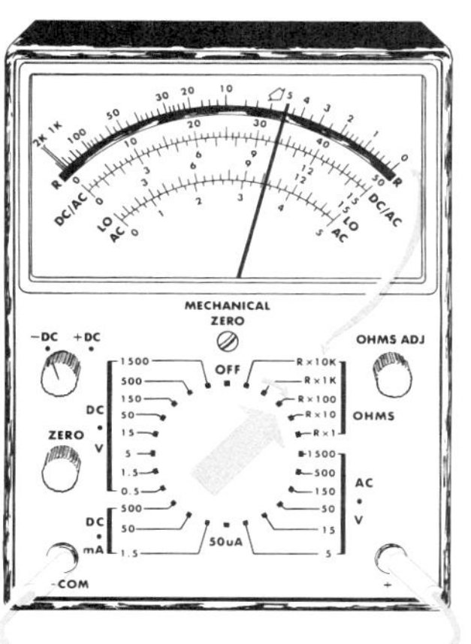

STEP 6. CONNECT THE RESISTOR TO BE MEASURED TO THE TWO TEST LEADS (−COM AND +).

STEP 7. SWITCH THE POINTER KNOB DOWN THROUGH THE OHMS RANGES UNTIL YOU GET THE NEEDLE SOMEWHERE ON THE HALF OF THE SCALE ON THE ZERO OHMS SIDE. YOU CAN GET A CORRECT READING ANYWHERE ON THE SCALE, BUT THE SCALE IS MORE CROWDED ON THE OTHER HALF OF THE SCALE. THE CROWDING MAKES IT HARD TO GET AN ACCURATE READING. ONCE IN A WHILE YOU WON'T BE ABLE TO FIND ANY SETTING ON THE POINTER KNOB THAT WILL KEEP THE NEEDLE OUT OF THE CROWDED SIDE OF THE SCALE. IN THAT CASE YOU WILL JUST HAVE TO DO THE BEST YOU CAN.

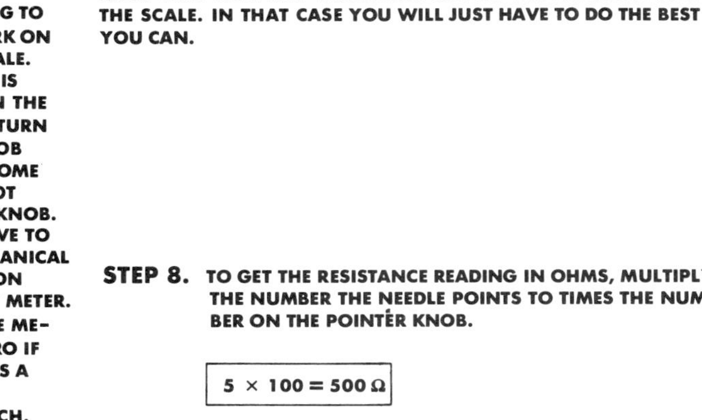

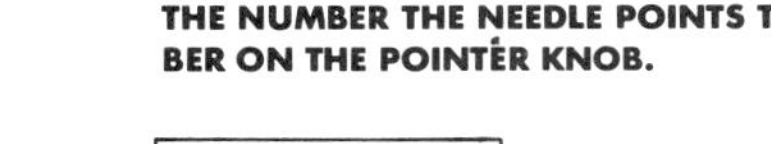

STEP 8. TO GET THE RESISTANCE READING IN OHMS, MULTIPLY THE NUMBER THE NEEDLE POINTS TO TIMES THE NUMBER ON THE POINTER KNOB.

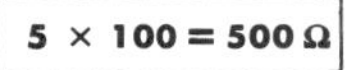

$5 \times 100 = 500\,\Omega$

Figure 23-4: Using a VOM to measure resistance.

SELF CHECK

Look at the meter, and read what the meter needle says with the pointer knob set at each of the four positions shown on the drawing.

1. ____________
2. ____________
3. ____________
4. ____________

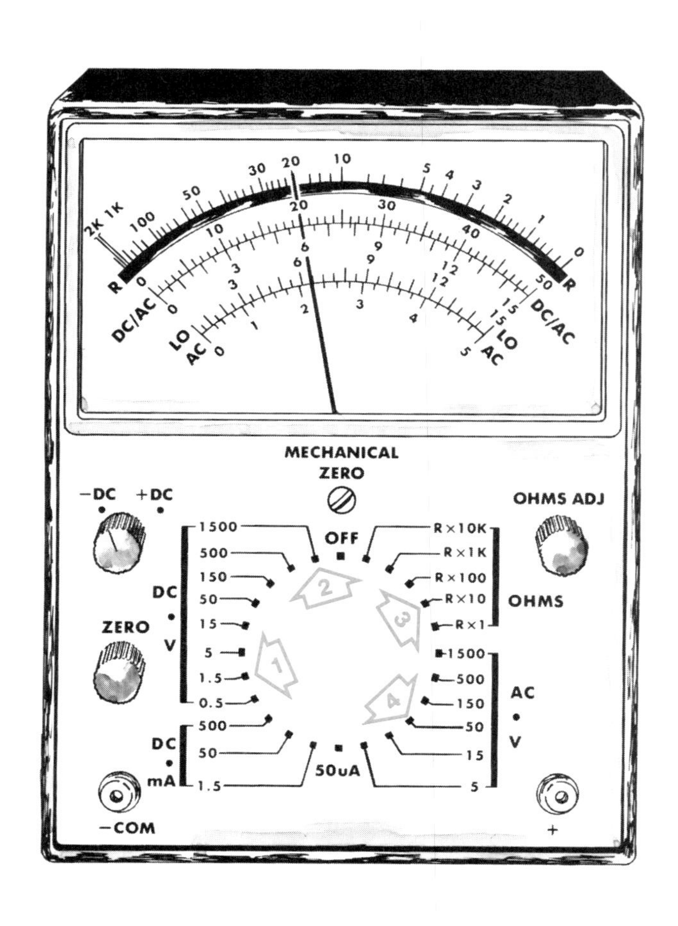

Figure 24-1: A charged capacitor stores electrical energy like a stretched spring stores mechanical energy.

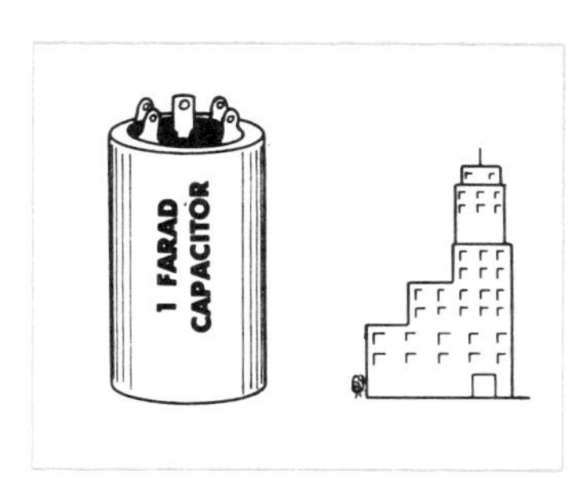

1 FARAD OF CAPACITY EQUALS A CHARGE OF ONE COULOMB OF ELECTRONS STORED — IF ONE VOLT POTENTIAL IS APPLIED.

Figure 24-2. A 1 Farad capacitor would be a giant. It could store enough energy to light a 100 watt light bulb for one second.

Capacitors are electrical storage tanks (figure 24-1). Simple capacitors are made of two plates separated by an insulating material. This insulating material is called a dielectric. Each of the two plates is capable of accepting and storing an electrical charge for a short period of time. Capacitors are not used to store large amounts of electricity. It would take a capacitor the size of a football field to store the energy that can be held in a small car battery.

The unit of capacitance is the farad. A one farad capacitor can store enough energy to light a 100 watt light bulb for about one second (figure 24-2). Most capacitors are much smaller than 1 farad. They are usually rated in microfarads and picofarads.

A microfarad is one millionth of a farad. The abbreviation for microfarads is μf. Other capacitors are rated in picofarads. A picofarad is one millionth of a microfarad. The abbreviation for picofarad is pf.

A capacitor is charged when a voltage source is used to move electrons from one of its plates to the other. One plate becomes very negative because it has a surplus of electrons. The plate that gave up its electrons becomes positive (figure 24-3).

Capacitors charge very quickly when they are connected to a voltage source. They will charge up to the same voltage as that of the voltage source and stop when they reach that voltage. When the capacitor is disconnected from the voltage source it will store that voltage

1. SWITCH OPEN—CAPACITOR HAS NO CHARGE.

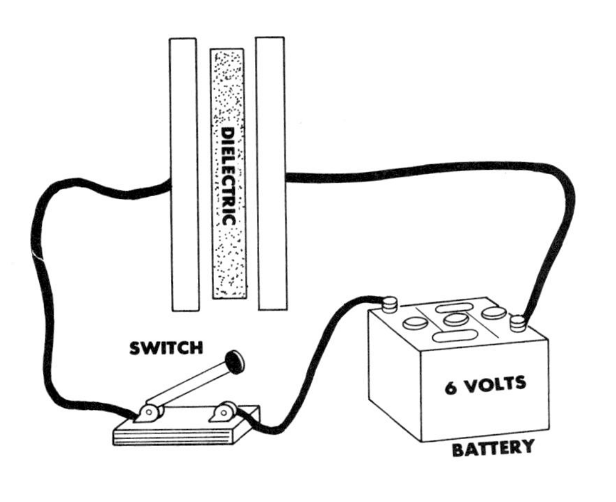

2. SWITCH CLOSED — THE BATTERY DRIVES ELECTRONS FROM THE POSITIVE (+) PLATE ONTO THE NEGATIVE PLATE.

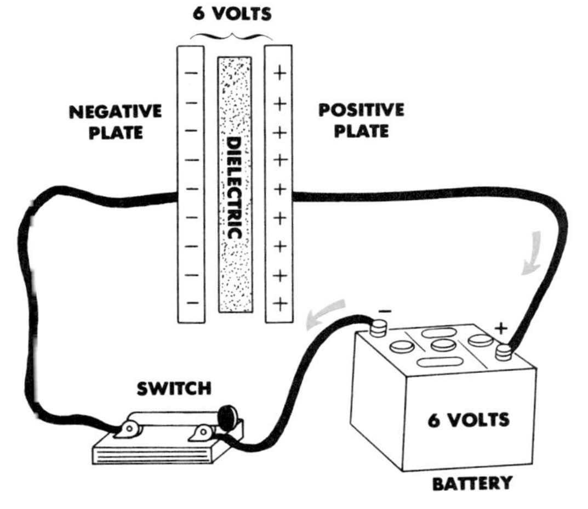

3. WHEN THE CAPACITOR IS DISCONNECTED, IT HOLDS THE 6 VOLT CHARGE. IT IS MUCH LIKE A 6 VOLT BATTERY.

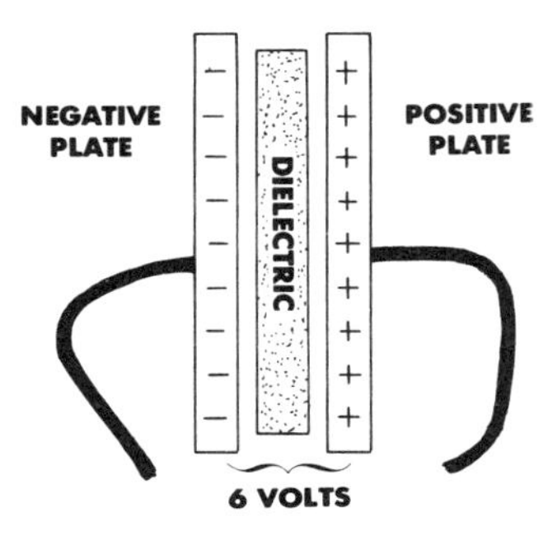

4. WHEN THE CAPACITOR WIRES ARE TOUCHED TOGETHER, THE EXTRA ELECTRONS ON THE NEGATIVE (−) PLATE RUSH BACK TO THE POSITIVE PLATE. THE CAPACITOR HAS BEEN DISCHARGED.

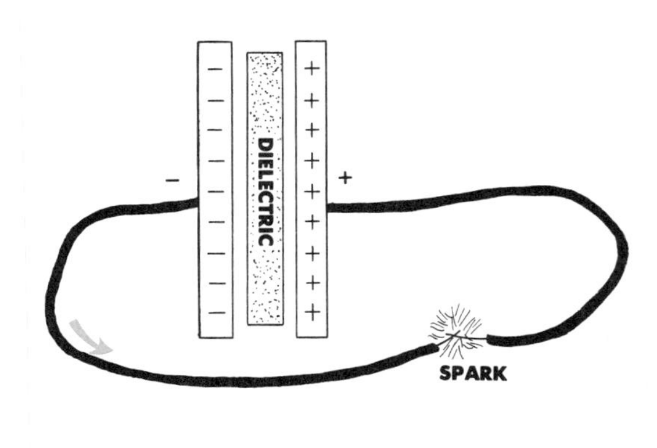

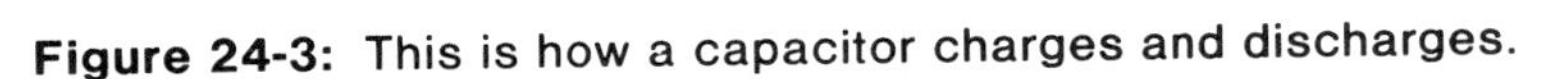

Figure 24-3: This is how a capacitor charges and discharges.

until a circuit is completed between the two plates. When a circuit is completed between the two plates the extra electrons will drain off the negative plate and the capacitor will be discharged.

The material that separates the two plates in a capacitor is called the dielectric (figure 25-4).

Capacitors have two ratings:

- Capacity in farads
- Voltage rating

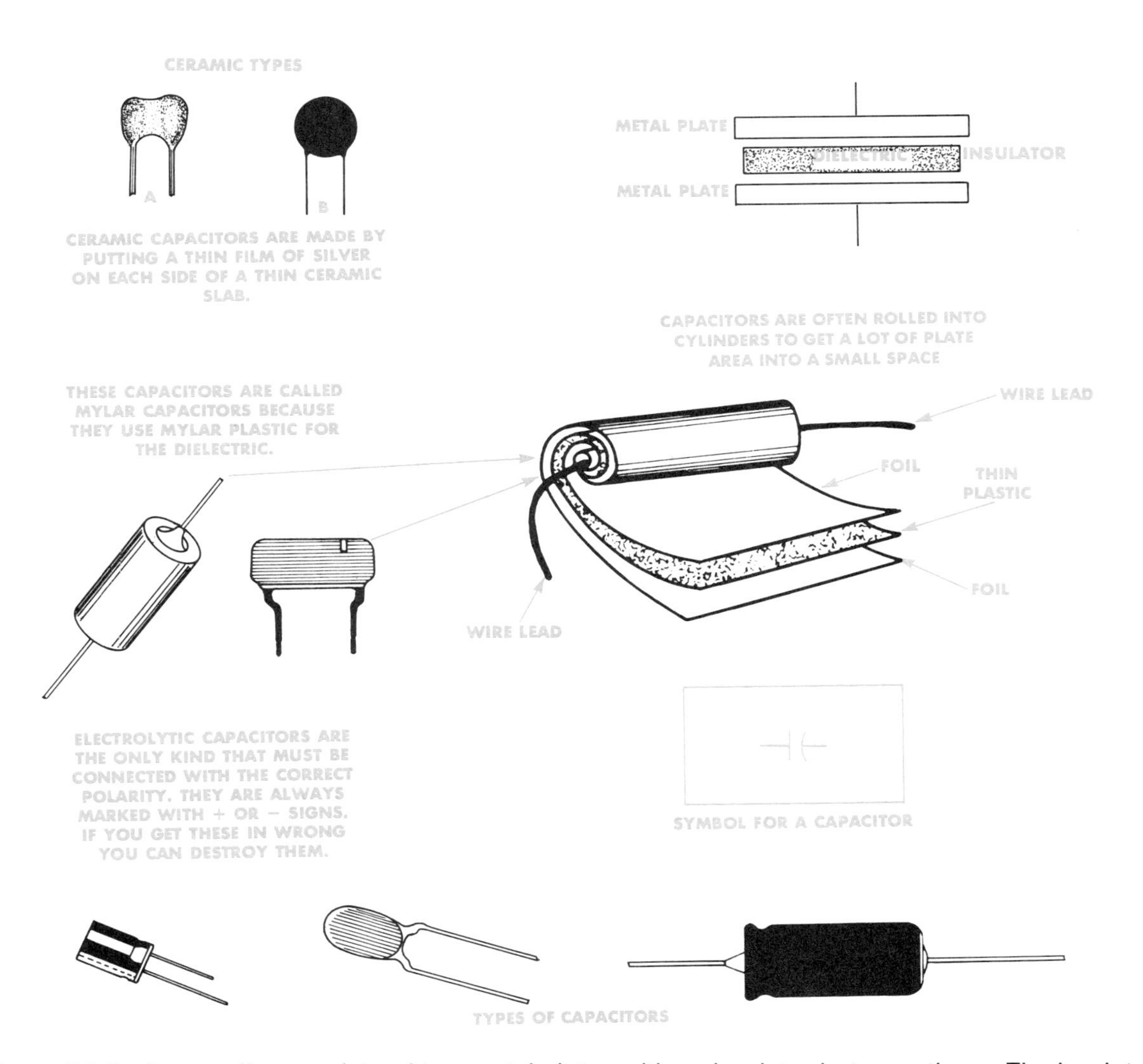

Figure 24-4: A capacitor consists of two metal plates with an insulator between them. The insulator is called the dielectric. Large plates and a thin dielectric make a larger capacitor.

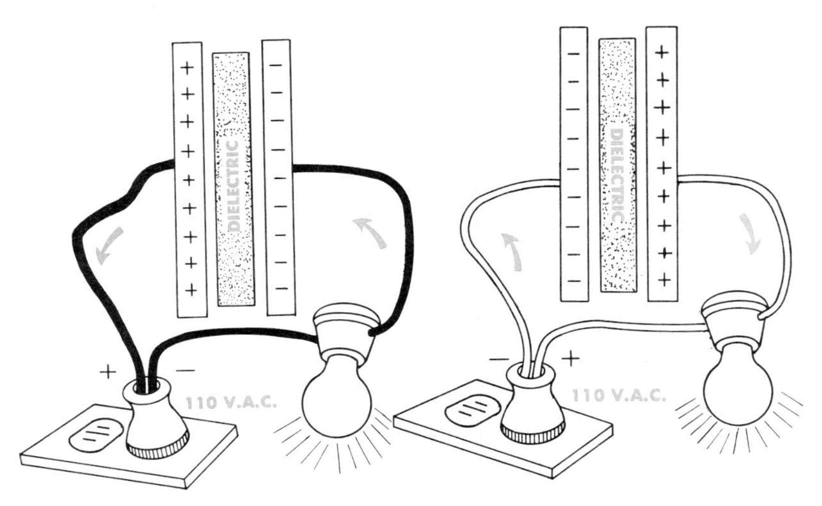

Figure 24-5: During the first one-half of an alternating current cycle current flows to charge one plate of the capacitor while the other plate discharges. The process is reversed during the second half of the alternating current cycle.

Always use a capacitor that has a voltage rating higher than the voltage of the circuit. If, for example, you are fixing a radio that is powered by a 9 volt battery, all the capacitors you use should have a voltage rating higher than 9 volts. A 50 volt capacitor can be used in a 10 or 25 volt circuit, but a capacitor rated for only 25 volts can not be used in a 50 volt circuit.

When a capacitor is used in an alternating current circuit it will charge in one direction as the voltage builds up. When the AC reaches its peak voltage and begins to drop off the capacitor will discharge until the voltage drops to zero. When the AC reverses itself the capacitor charges and discharges in the opposite direction with the opposite polarity (figure 24-5).

SELF CHECK

1. What is a capacitor made of?
2. Is the dielectric an insulator or a conductor?
3. Which kind of capacitor has to be hooked up in a certain way?
4. What is a microfarad?

RESISTOR-CAPACITOR (RC) TIMING CIRCUITS — Unit 25

Capacitors are used in many ways, but one of their most common uses is to control timing circuits. Photographic enlarger timers and burglar alarm delays are examples of devices that use capacitors to time their operation (figure 25-1).

When a capacitor is connected to a battery, electrons flow from the battery to the plate of the capacitor, and it charges almost instantly. As soon as the capacitor is completely charged the current stops flowing. We can control the time required to charge a capacitor by adding a resistor in series with the capacitor.

A large value capacitor connected in series with a high value resistor will

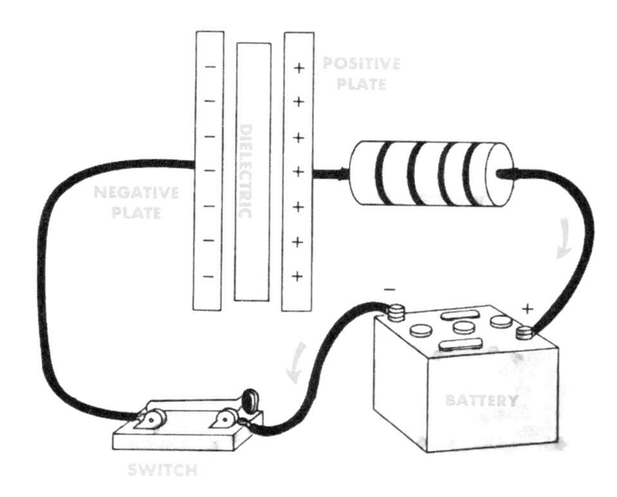

Figure 25-2: A resistor in series with a capacitor reduces the current charging the capacitor. We can make the capacitor take as long to charge as we desire.

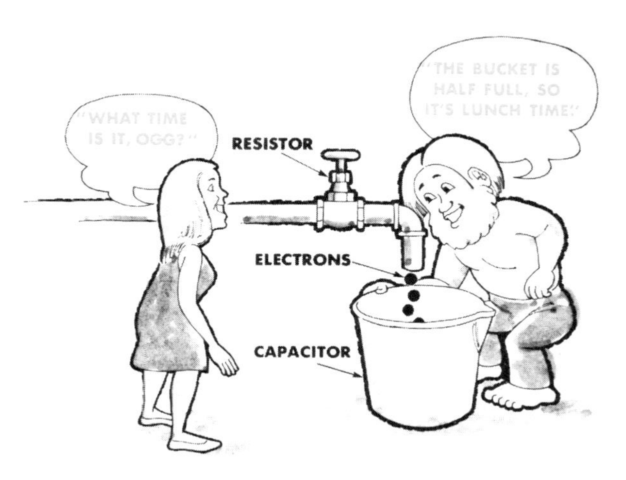

Figure 25-1: Our ancestors didn't have resistors and capacitors to help them make a timing circuit. They used other devices, but the principles are the same.

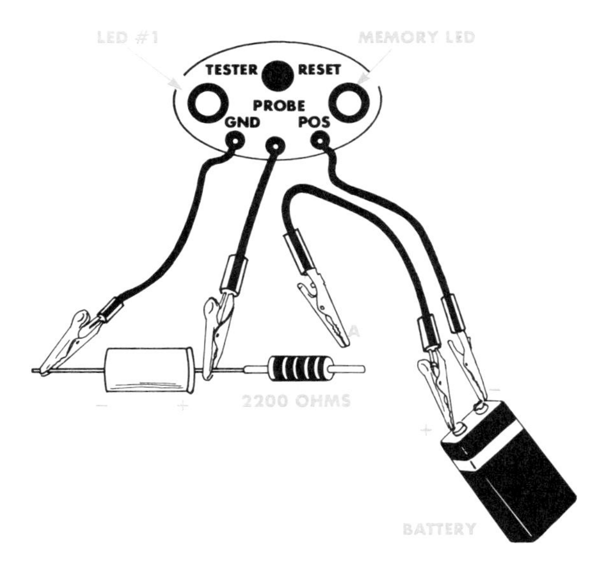

Figure 25-3: You can see how a timing circuit works on your tester.

give us a long charging time. A low value resistor connected in series with a small value capacitor will give us a short charging time (figure 25-2). It takes the same amount of time to discharge a capacitor through a resistor as it takes to charge it.

We can make a simple resistor-capacitor timing circuit by connecting a resistor and a capacitor to the tester we built in Unit 7. Just hook it up as shown in figure 25-3. Changing the values of the resistor and/or the capacitor will change the time required to light the LED. Some typical resistor-capacitor timing combinations are shown in figure 25-4. Try as many as you can.

The following formula can be used to figure out any timing combination you might need:

Charge Time Equals
Resistance Times Capacitance

To find the charge time of a resistor-capacitor (RC) circuit (in seconds), simply multiply the value of the resistor (in megohms) by the value of the capacitor (in microfarads).

The formula (Charge Time = Resistance × Capacitance) is called the Time Constant Formula. It takes 5 time constant periods to completely charge a capacitor (figure 25-5).

1 TIME CONSTANT TIME IN SECONDS	CAPACITOR	RESISTOR (OHMS)
1	10 µF	100,000
2	10 µF	200,000
5	50 µF	100,000
10	100 µF	100,000

Figure 25-4: Try these capacitor and resistor values with your tester.

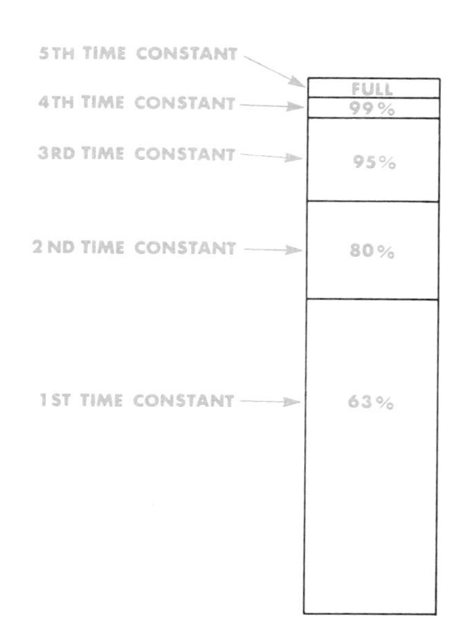

Figure 25-5: It takes five time constant periods to completely charge a capacitor.

SELF CHECK

1. True or false: A large capacitor and a high value resistor requires a short time to charge.
2. A timing circuit has a 10 microfarad capacitor and a 1 megohm resistor. How many seconds is one time constant?
3. True or false: A resistor can be used to slow down the charging rate of a capacitor.
4. How fast will a capacitor charge without any resistance in the circuit?

In the unit on generators you learned that a magnet moving near a coil causes a current to flow.

What is really important is that the magnetic field be in motion. With a permanent magnet we can't, of course, move the field without moving the magnet.

With an electromagnet, we can make the field expand and collapse by changing the electrical current through it. A field that is expanding and collapsing has to be in motion.

So, what happens if we put a coil next to the electromagnet with its field expanding and collapsing? The extra coil finds itself in a moving magnetic field. It doesn't know what makes the field move. It just obeys the law and a current flows through it just as it would in a generator.

In figure 26-1 the lamp is lit but there is no electrical connection to the battery that it gets its power from.

The light can only stay lit as long as somebody is wiggling the shaft on the potentiometer to keep the current changing.

What we need is a type of electrical current that is always changing by itself.

If we can find that kind of current we can get rid of the battery and potentrometer. And we won't need anyone to wiggle the shaft.

What we need is alternating current.

When we use coils in this way we call them a transformer. One coil is called

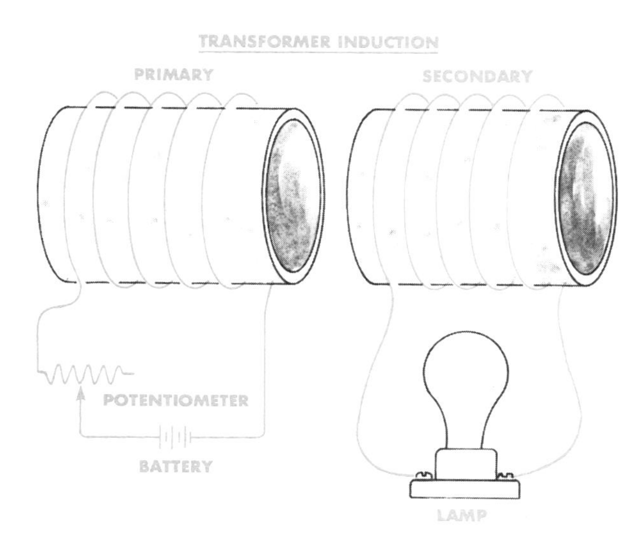

Figure 26-1: Putting a second coil in the expanding or contracting field of a transformer causes current to flow in the second coil.

the primary and the other is called the secondary.

Transformers come in a number of shapes and sizes (figure 26-2). Most of them have some kind of iron core. The core is often laminated to help keep the electrical losses down.

Transformers are used with alternating current to step voltages up or down. We can get twice as much voltage out of the secondary as we are putting into the primary. All we have to do is put twice as many turns of wire in the secondary winding. If we want half the primary voltage from the secondary we use half as many turns in the secondary.

We never get something for nothing.

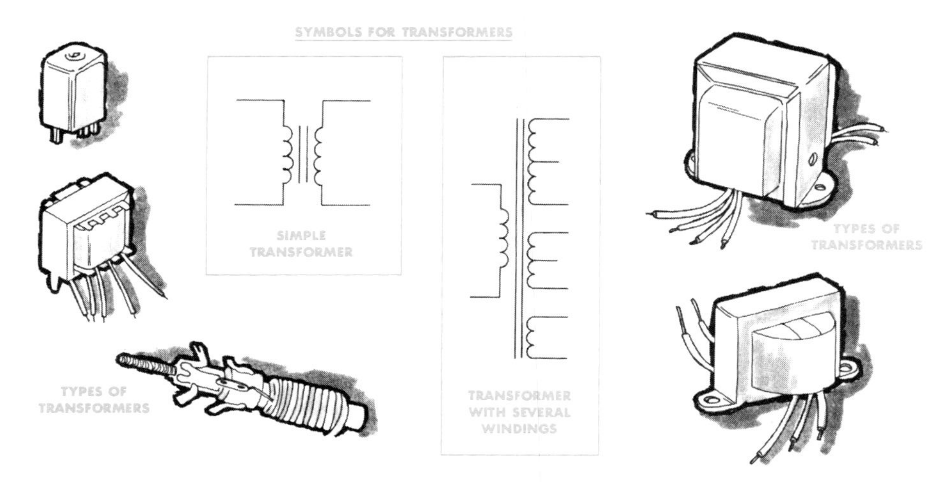

Figure 26-2: Transformers can be very simple or very complicated. They come in many sizes and shapes.

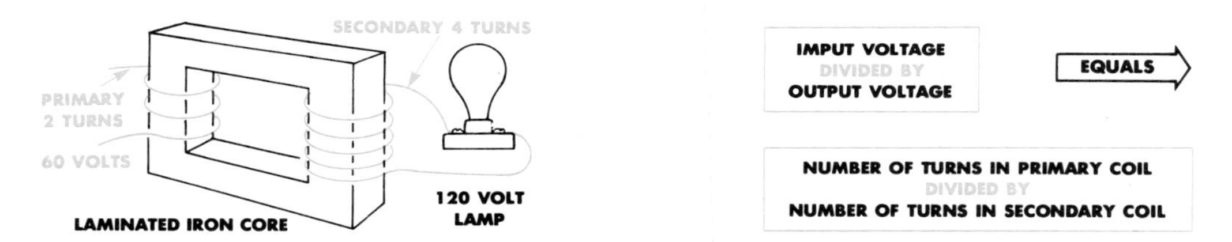

Figure 26-3: A transformer will step AC voltages up or down. How much the voltage changes is determined by how many turns of wire there are in the secondary and primary coils.

If we step up the voltage we must give up an equal amount of current. The power in must equal the power out. Volts × amps in the primary always equals volts × amps in the secondary. In figure 26-3 the current in the primary will have to be twice the current in the secondary. Suppose the lamp in the secondary uses 1 amp at 120 volts. The secondary power is 1 × 120 = 120 watts. If we have 60 volts in the primary there will have to be 2 amps of primary current. 60 volts × 2 amps = 120 watts.

SELF CHECK

1. What kind of current is a transformer usually used with?
2. A transformer has 10 primary turns and 100 turns on the secondary. 10 volts AC is connected to the primary. What is the voltage at the secondary?
3. Can a transformer be used to step voltage up?
4. Can a transformer be used to step voltage down?

A capacitor stores electrical energy like a stretched spring stores mechanical energy.

A coil stores energy also, but the energy stored in a coil acts more like the energy stored in a weight lifted off the floor. If you drop the weight to the floor, the energy stored in it will be used up.

Gravity does not have to be involved. If you have ever been hit by a flying baseball you know how much energy can be stored in a moving object.

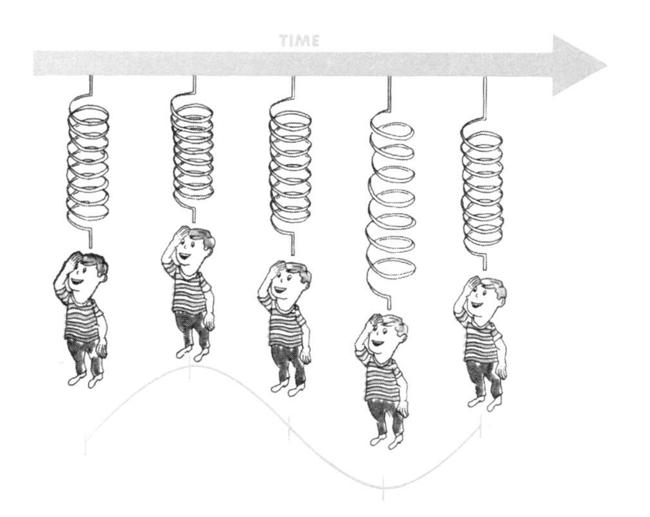

Figure 27-1: An electrical cycle is one complete reversal of an alternating current from zero through positive, back to zero, through negative, and back to zero again. Cycles are measured by the number of times per second they complete this sequence. If an alternating current completed its sequence 60 times each second, we would say it was a 60 hertz current.

If we take a spring and a weight as shown in figure 27-1 and give the weight a start, it will keep going up and down for a while.

If we put a pen on the weight and move the weight and spring along at a steady rate the pen will trace out a sine wave curve. This is exactly the same as the sine wave curve we get from an alternating current generator.

When we pull the spring down to start it, energy is stored in the spring. When we let go, the spring pulls the weight up. When the spring is all the way up it stops pulling up and the weight falls. This movement is called oscillation.

Let's look at a different type of mechanical oscillator and its equivalent in electrical parts (figure 27-2). If we start the weight on the mechanical oscillator moving, the spring will store energy until it is stronger than the movement of the weight. The spring will then force the weight back in the other direction until the sequence is repeated again.

If we start an electronic oscillator by charging the capacitor, it will discharge through the coil. The discharge current through the coil stores energy in the coil's magnetic field. When the capacitor is empty, the coil's magnetic field will collapse and drive electrons back into the capacitor. The capacitor will then discharge through the coil, starting the cycle again.

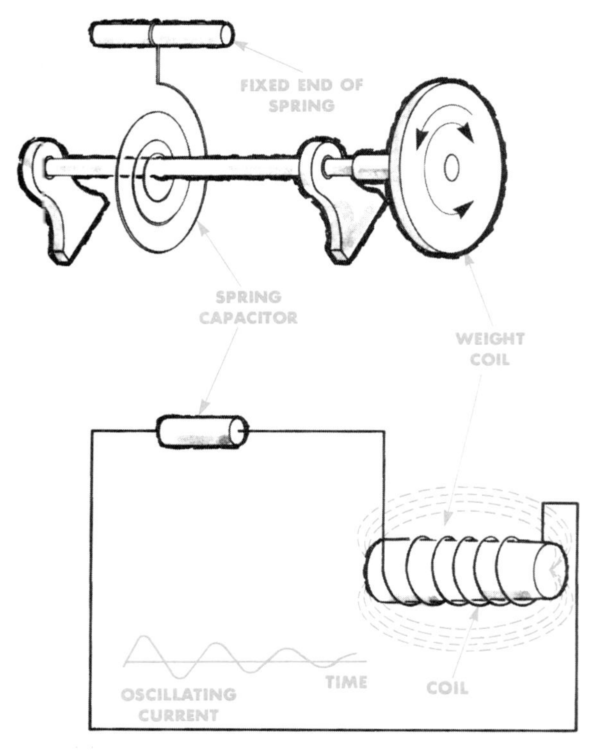

Figure 27-2: A mechanical oscillator and an electronic oscillator. Notice their similarities.

The frequency of oscillation (in Hertz) is called the resonant frequency. Resonant oscillators would go on forever if it weren't for losses due to friction or resistance.

Coil and capacitance resonant circuits are used to control the frequency of radio and television waves, tones for electronic music, and hundreds of other things.

Every combination of coil and capacitor has its own particular resonant frequency. A particular combination will oscillate only at its particular resonant frequency.

- An electronic resonant circuit is made up of a coil and a capacitor.
- A resonant circuit would oscillate forever if there were no friction.
- A particular resonant circuit will oscillate at one and only one frequency.

SELF CHECK

1. What kind of wave is generated by a resonant circuit?
2. True or false: Each combination of coil and capacitor has several resonant frequencies.
3. What makes a resonant oscillation slowly coast to a stop?
4. What kind of wave does a resonant circuit produce?

AMPLIFIERS — Unit 28

Electricity became electronics with the invention of amplification. We used to talk about electronics as a branch of electricity. Now it's the other way around.

The first electrical amplifier was the relay. It was used by the telegraph industry to amplify the Morse Code signals when they got too weak to work the telegraph receivers.

Modern transistor amplifiers are used to amplify radio and television signals that are too weak to move a loudspeaker or make images on a picture tube screen. Modern transistor amplifiers are much more complicated than the relay amplifier, but they both work on the same principle.

Figure 28-2 shows a relay amplifier circuit. When the wires A and B are touched together current flows through the coil closing the contacts. The closed contacts complete the 120 volt circuit and light the light bulb.

Figure 28-1: When we press our foot down on the accelerator of a car, we are using a little power to control a lot of power (horsepower). This is amplification. The force pushing down the pedal is the input and the force turning the wheels is the output. Whatever else the automobile is it is also a mechanical amplifier.

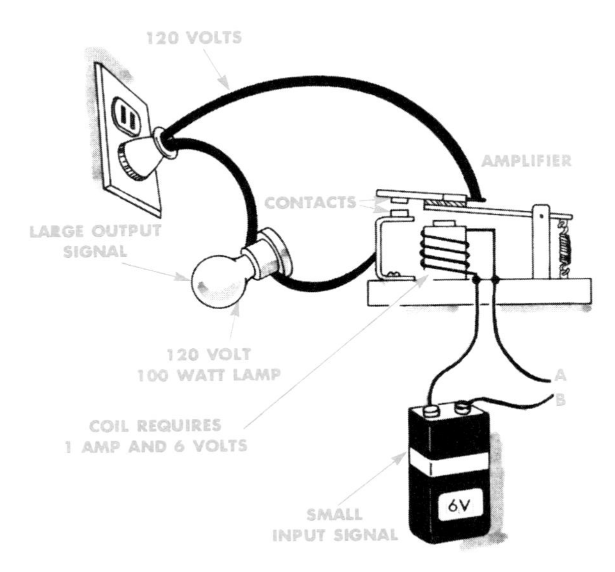

Figure 28-2: The relay was the first electrical amplifier.

The word "gain" with a number tells how many times bigger the output signal is than the input signal. We can get three kinds of gain from an amplifier.

- Voltage gain
- Power gain
- Current gain

Voltage gain is one type of amplifier gain. The formula for finding voltage gain is:

Voltage Gain Equals
Output Voltage ÷ Input Voltage.

In our relay amplifier, the input voltage is 6 volts and the output voltage is 120 volts. So:

Voltage gain = 120V ÷ 6V = 20.

The voltage gain of the relay amplifier is 20. This is just a relative number, it does not represent volts.

Power gain is the second type of amplifier gain. The formula for finding power gain is:

Power Gain Equals
Output Power ÷ Input Power.

The output power of out relay amplifier is 100 watts. The input power is 6 watts (Input power = volts × amps, and 1 amp × 6 volts = 6 watts). The power gain in our relay amplifier equals 16.6.

Power gain = 100 watts ÷ 6 watts = 16.6.

Current gain is the third type of gain an amplifier can have. The formula for finding current gain is:

Current Gain Equals
Output Current ÷ Input Current.

Our relay amplifier doesn't have much current gain, but other amplifiers might have high current gain coupled with low voltage gain.

There are two basic kinds of amplifiers.

- Digital amplifiers
- Analog amplifiers

Digital amplifiers are simple on-off amplifiers like our relay amplifiers. They are used in computers and industrial control system.

Analog amplifiers amplify a changing input signal into a changing output signal. They are used for amplifying sound, radio waves, TV pictures and so on. An example of an analog amplifier is shown in figure 29-3.

Changing the input signal will vary the output signal. A change of from zero to 6 volts at the input coil could vary the output voltage from zero to 120 volts. The brightness of the lamp

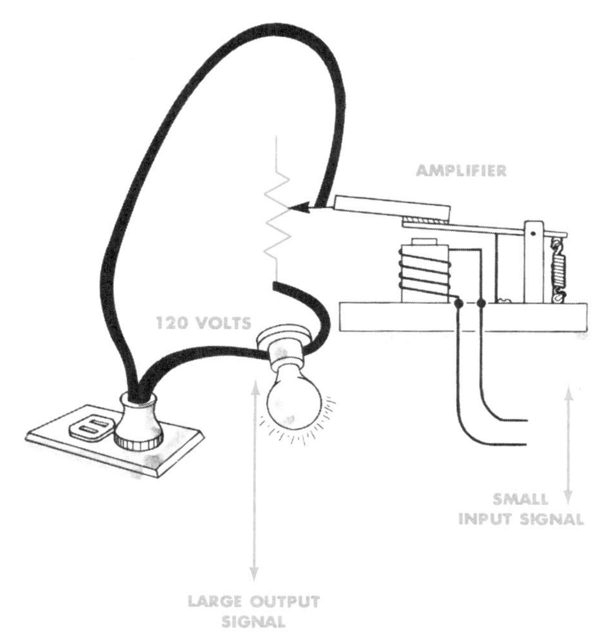

Figure 28-3: An Analog amplifier can amplify a changing signal.

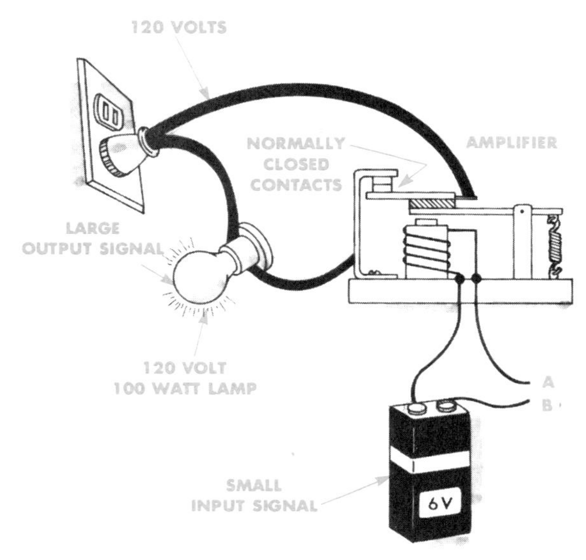

Figure 28-4: An inverting amplifier produces a high output voltage when the input voltage is low. It produces a low output voltage when there is a high input voltage.

would adjust from no light at all, when the input voltage is zero, to full on, when the input voltage is 6 volts. But, the voltage gain would still be the same. For example, 3 volts on the input coil would deliver 60 volts to the lamp. Voltage gain = 60 ÷ 3 = 20.

Both digital and analog amplifiers are available in two versions.

- Inverting amplifiers
- Non-inverting amplifiers

The lamp is the output of our relay amplifier. For it to be on, the input must be on. When the input is off, the light turns off (off=off). This is called a non-inverting amplifier. The amplifier in figure 28-4 is an inverting amplifier. When the input is on, the light (output) is off (on=off). The light turns on when the signal is off. The light gets dim when the input voltage is high. It gets bright when the input voltage is low.

SELF CHECK

1. What is the purpose of an amplifier?
2. What are the three kinds of amplifier gain?
3. What is the difference between a digital amplifier and an analog amplifier?
4. Which type of amplifier produces a changing output signal?

Unit 29 DIODES AND TRANSISTORS

Most diodes and transistors are made from silicon crystal. Pure silicon is a poor conductor of electricity. Impurities must be added to the pure silicon to improve its ability to conduct electricity for diodes and transistors to work.

Two different kinds of impurities are used to improve the conductivity of the pure silicon used in diodes and transistors. They are:

- "N" type impurities
- "P" type impurities

"N" type impurities add a lot of extra electrons to the silicon crystal. The "N" stands for negative, because anytime we have a lot of extra electrons concentrated in one spot we have a "negative" charge.

"P" type impurities have a deficiency of free electrons. The "P" stands for positive, because anytime we have a deficiency of free electrons concentrated in one spot we have a "positive" charge. This deficiency of electrons in the silicon provides a number of free paths through which electrons entering from outside the crystal can flow. These free paths are called holes. Holes behave like electrons with a positive charge.

The diode (figure 29-1) is a one-way electronic valve. Current can flow from its cathode (negative side) to its anode (positive side), but it can not flow from its anode (+) to its cathode (−).

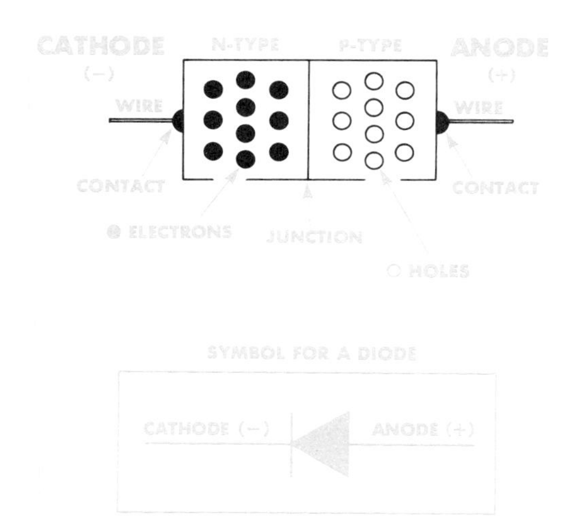

Figure 29-1: The diode is a one-way electronic value.

Diodes are made by melting "P" type and "N" type crystals together.

When a diode is made, the two types of silicon crystal are melted together at a temperature of 1000 degrees or so. As the diode cools some of the free electrons from the "N" side move permanently into some of the holes on the "P" side. This action creates a kind of electrical wall called the "junction" (figure 29-2).

Electrons cannot flow through this wall unless the voltage pushing them is at least .6 volts. This is true of all silicon junctions.

Light emitting diodes (LED'S) need about 1.5 volts before they can conduct,

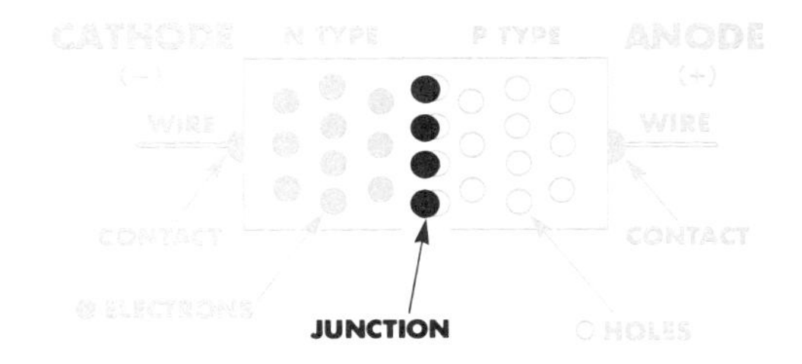

Figure 29-2: The diode will not conduct with the battery connected this way. This is called reverse bias.

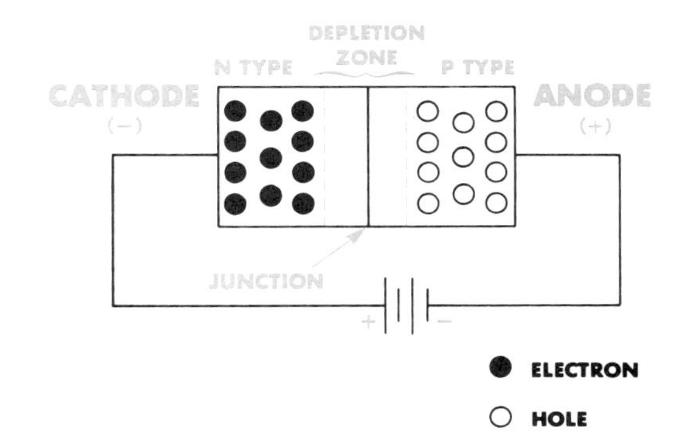

Figure 29-3: The diode will conduct when the battery is connected this way. This is called forward bias.

but they are not made of silicon. They are made of gallium arsenide.

A diode works this way because of the electrical law:

- Unlike charges attract each other
- Like charges repel each other

The positive pole of the battery attracts electrons away from the junction toward the end of the crystal that is connected to the positive pole of the battery. Since holes behave as if they have a positive charge, they are attracted to the negative pole of the battery. They move away from the junction toward the end of the crystal connected to the negative pole of the battery. All the electrons and holes have now moved away from the junction leaving a space between the "P" and "N" crystals in which there are no free electrons or holes. This space is called the "depletion zone" (figure 29-3). The depletion zone is an insulator. It is an insulator because all of the free electrons and all of the holes have been swept out. No current can flow.

Since electrons flow from negative to positive, and a diode won't let current flow from its anode (positive side) to its

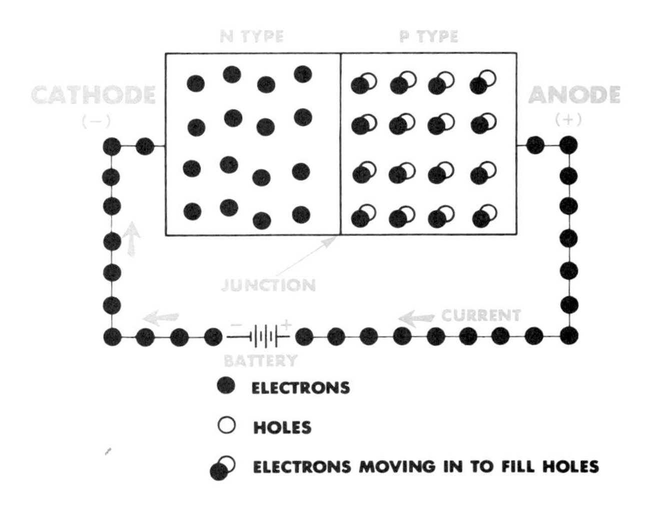

Figure 29-4: Current flows through a diode because the negative pole of a battery repels electrons toward the junction while the positive pole also repels holes toward the junction. This completes the circuit.

cathode (negative side), no current can flow in a circuit when the negative pole of a battery is connected to the anode of a diode as shown in figure 29-3. This is called reverse bias.

If we reverse the battery connections as shown in figure 29-4, current can flow through the diode. It can flow because the negative pole of the battery repels electrons toward the junction, and the positive pole of the battery repels the positivitly charged holes toward the junction.

This condition allows electrons to hop from positivly charged hole to positivly charged hole until they reach the wire that takes them back to the battery. This is called forward bias.

The transistor (figure 29-5) is an electronic amplifying device. It is made of silicon just like a diode. However, a transistor has two junctions instead of one.

There are two kinds of transistors:

- PNP
- NPN

"PNP" means positive, negative, positive. A "PNP" transistor is made by sandwiching a layer of silicon with "N" type impurities between two layers of silicon with "P" type impurities (figure 29-6).

"NPN" means negative, positive, negative. An "NPN" transistor is just

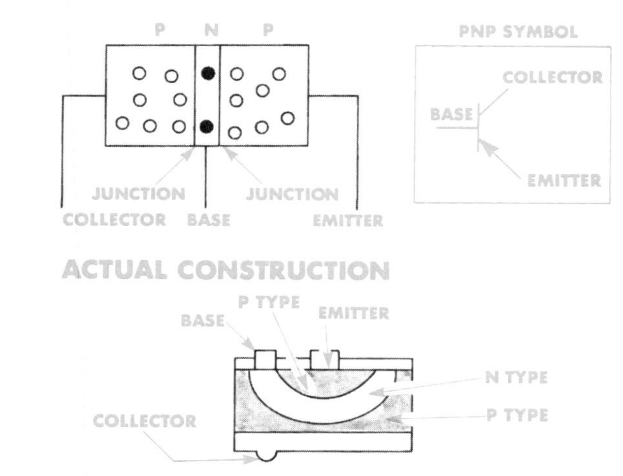

Figure 29-6: The "PNP" transistor.

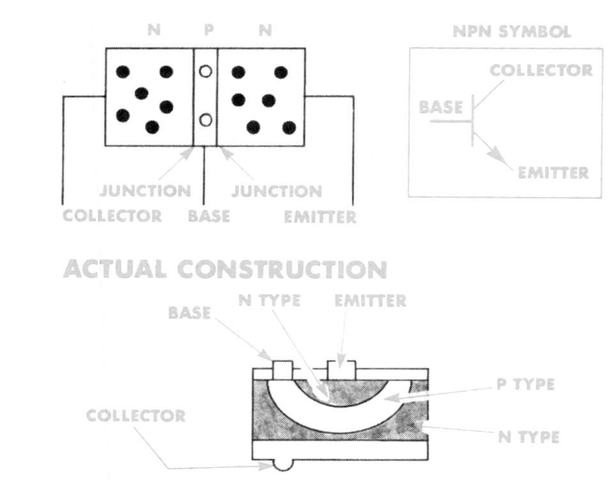

Figure 29-7: The "NPN" transistor.

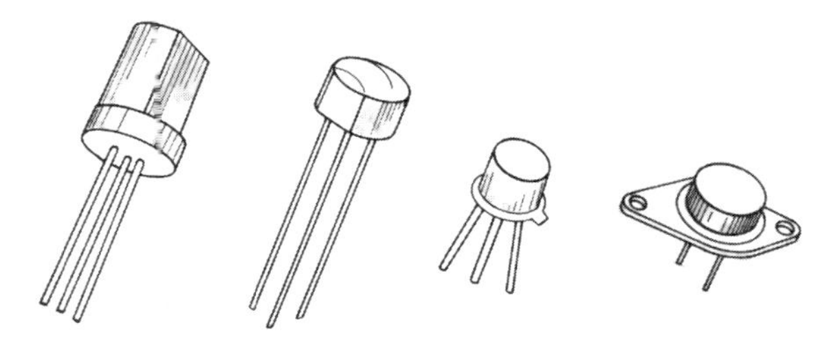

Figure 29-5: Some transistor case styles.

the opposite of a "PNP" transistor (figure 29-7).

Transistors have three parts:

- Emitter
- Collector
- Base

The emitter emits electrons. The collector collects electrons. The base controls electron flow.

Transistors are very complicated so you will have to use your imagination to help with this simple explanation.

The transistor in figure 29-8 is an "NPN" type. There are lots of electrons in both the emitter and the collector blocks. The base is very thin and has only a few stepping stones (holes) leading to the "base to emitter" subway. If we apply a small positive voltage to the base, electrons will flow from the emitter to the base just as if the transistor was a diode. The first few electrons that are pulled from the emitter to the base by the positive charge of the battery will block open the gates. These electrons complete their circuit by passing through the "local" subway from the base back to the emitter.

Opening the gates allows thousands of electrons to rush through the base and on into the collector. These electrons complete their circuit by taking the "express" subway from the collector to the emitter. Along their way they can do things like operate a loud-

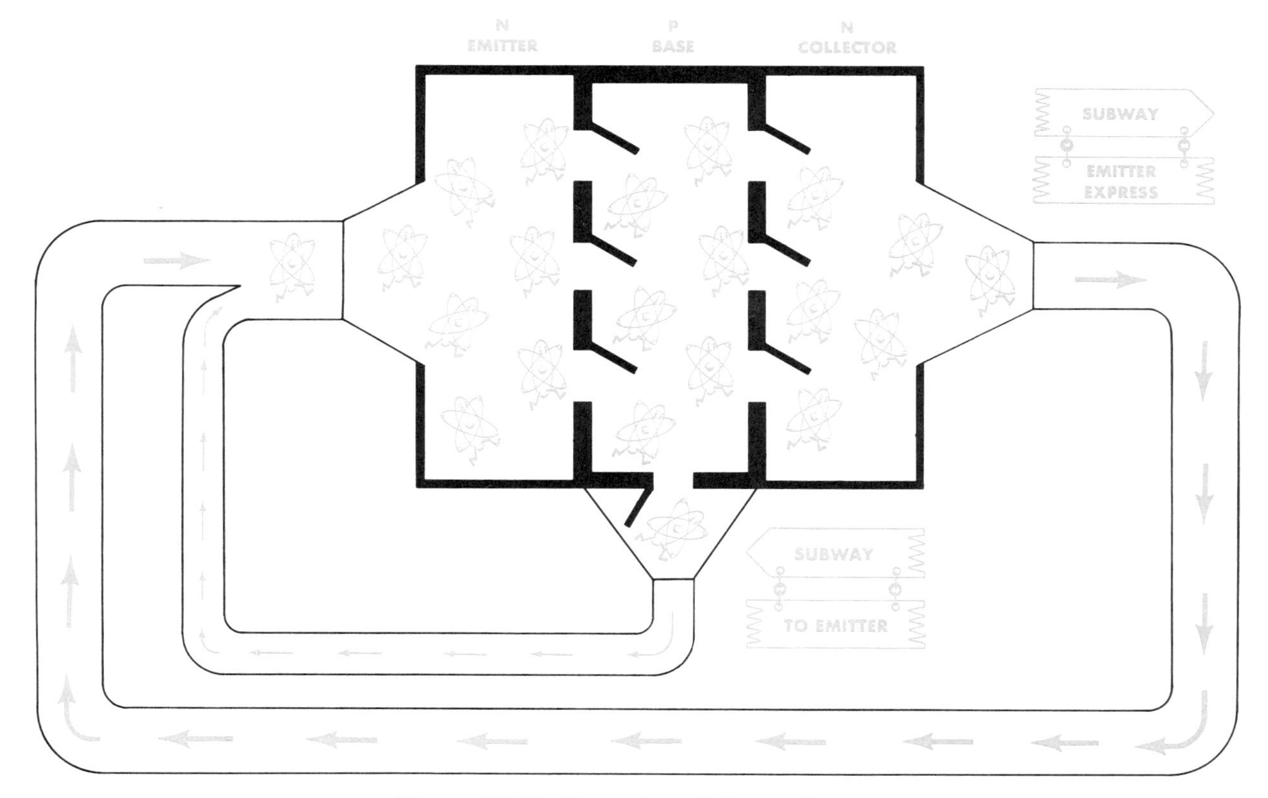

Figure 29-8: How a transistor works.

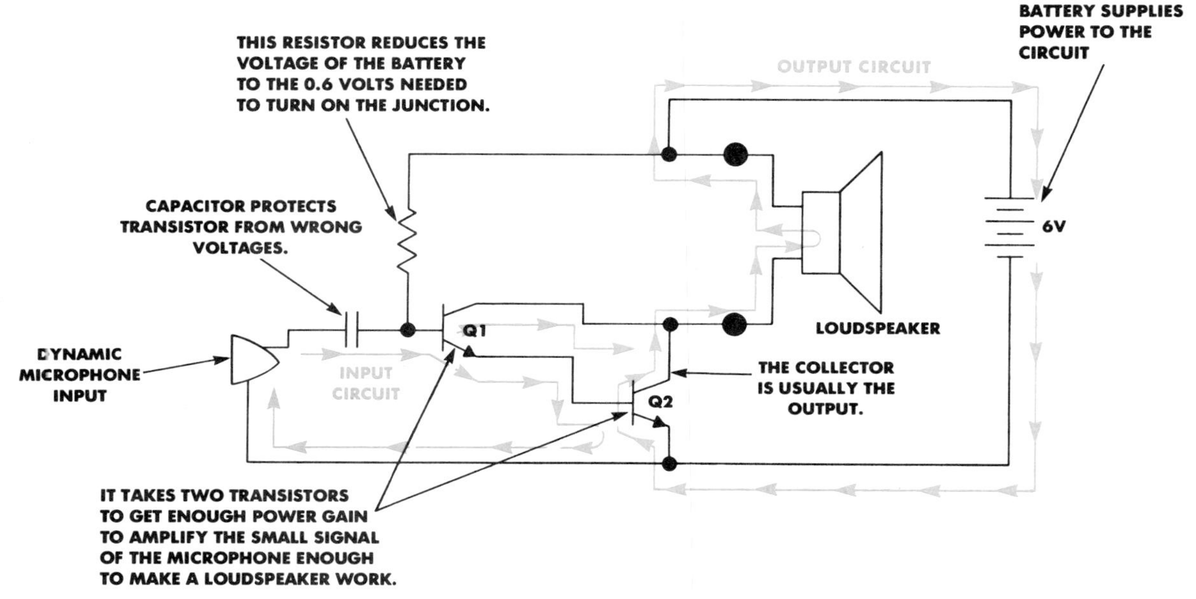

Figure 29-9: This is how a simple transistor amplifier works.

speaker or turn on a light. A small base voltage may open one gate. A bigger base voltage will open two gates and so on. The result is the same thing that happened with the relay amplifier we studied in Unit 26. For each electron that opens a gate a hundred more can follow it through. A small base current has resulted in a high collector to emitter current, and we have amplification.

The transistor can provide current, voltage, and power gain. A simple transistor amplifier is explained in figure 29-9. Other transistor amplifiers may have more parts, but they all work the same way.

SELF CHECK

1. True or False: A hole has a positive charge.
2. True or False: A diode can conduct electricity when it is reverse biased.
3. True or False: The anode of a diode must be connected to the positive pole of the battery to make current flow through it.
4. True or False: The transistor can provide voltage gain, current gain and power gain.

The fellow in the cartoon (figure 30-1) isn't kidding. Figure 30-2 is the drawing of a stereo amplifier that you can buy for less than $5.00. All of the 60 transistors and other parts are on a tiny silicon chip about 1/8 inch square, and a few thousandths of an inch thick. All it needs is a battery, a record player or tuner, and a couple of speakers and it's ready to go.

It is hard to believe that it is possible to put all those parts on such a tiny chip, but it's true.

Analog amplifiers are all similar, but different ones operate best over certain frequency ranges.

All analog amplifiers have very high gains. Most of them have two inputs, one inverting and one non-inverting. There are also special analog integrated circuits that take the place of most of the parts in a radio or TV sound system.

General purpose analog IC amplifiers are called operational amplifiers (Op-amps). They have the special symbol shown in figure 30-3.

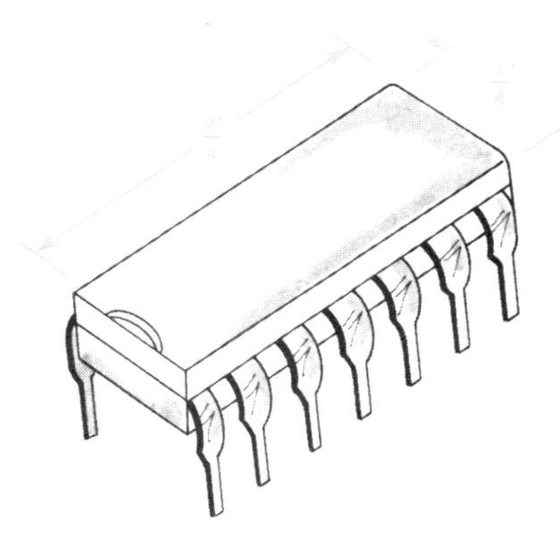

Figure 30-2: This is an integrated circuit stereo amplifier. It has 60 transistors, 49 resistors, 6 capacitors, and 3 diodes on a silicon chip 1/8 of an inch square.

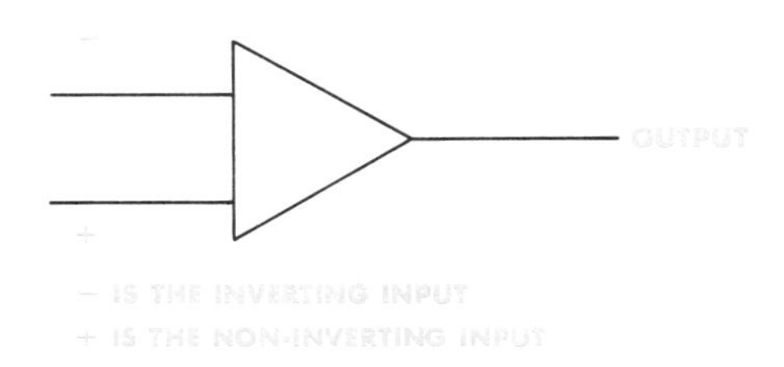

Figure 30-3: The special symbol for an Operational Amplifier.

Figure 30-1: Integrated circuits are called "chips" because they are so small.

Here is a list of only a few of the analog integrated circuits that are available:

- Op-amps
- Complete audio amplifiers, monaural and stereo from 1-50 watts
- Standard broadcast radio
- AM-FM radio
- Set of television building blocks
- Electronic organ and synthesizer tone generator
- AM-FM single side band detector for CB radios etc.

Only the OP-amp has a special symbol.

Figure 30-4 is the schematic diagram of a radio reciever made from two integrated circuits. Some coils, transformers, and capacitors must be used with the IC. Up until recently it has been impossible to put coils and large capacitors on the IC chip. Future IC's will have them on the chip.

SELF CHECK

1. Draw the symbol for an OP-amp.
2. Do OP-amps have both inverting and non-inverting inputs?
3. Give two examples of special analog integrated circuits that are available.
4. Name two electronic parts that could not be put on an IC chip until recently.

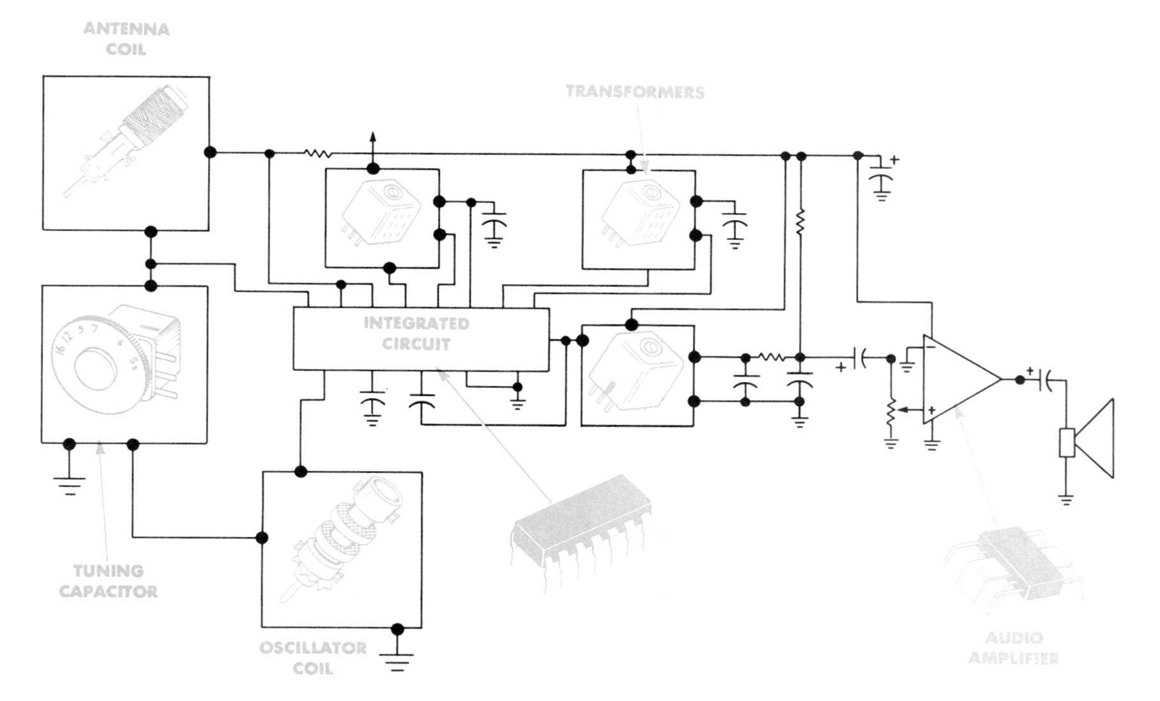

Figure 30-4: A radio receiver made from two integrated circuits.

INTEGRATED CIRCUITS (DIGITAL) — Unit 31

Digital circuits are on or off circuits. Because there are only two possible conditions we need a number system with only two values.

Digital computers compute and digital counters count. But they do it with binary numbers instead of our familiar decimal numbers.

We count by 10's. Digital systems count by 2's. We use digits 0-9. Digital systems use only 0 and 1.

Figure 31-1 shows how binary numbers work.

Every digital system from simple circuits to complete computers are built from gates. Gates are small building blocks that can be put together to make larger blocks. Actually, gates are like electronic relays that use +5 volts to stand for the digit 1 and 0 volts to stand for the digit 0.

DECIMAL NUMBER	BINARY NUMBER	8	4	2	1	
0	0000	○	○	○	○	0 + 0 + 0 + 0 = 0
1	0001	○	○	○	●	1 = 1
2	0010	○	○	●	○	2 = 2
3	0011	○	○	●	●	2 + 1 = 3
4	0100	○	●	○	○	4 = 4
5	0101	○	●	○	●	4 + 1 = 5
6	0110	○	●	●	○	4 + 2 = 6
7	0111	○	●	●	●	4 + 2 + 1 =7
8	1000	●	○	○	○	8 = 8

Figure 31-1: How the binary number system works.

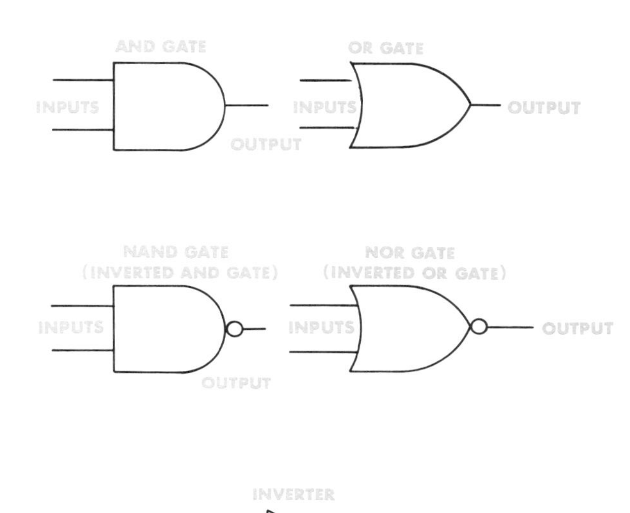

Figure 31-2: Here are the basic logic gates.

An electronic oscillator called a clock is used to produce a string of signals. These clock signals tell all of the gates in the system when it is time to turn on and off.

There are only 5 kinds of gates. With enough of these 5 kinds of gates put together in the right combination, we can build a giant computer.

A flip-flop is the simplest binary memory element.

We can take a few gates and hook them up to make a flip-flop. Figure 31-3 shows how the gates are connected to make a flip-flop.

Figure 31-4 shows a model of a flip-flop. The clock input makes the switch

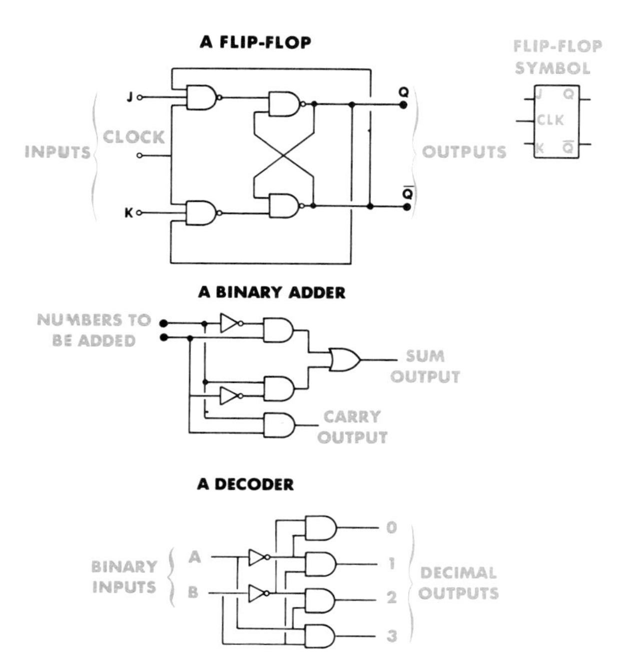

Figure 31-3: Gates can be connected to make a Flip-flop, a Binary adder, or a Decoder.

change positions at the right time. The flip-flop has two outputs. one is called Q, the other is called Q̄. Q̄ means not Q. Only one of the outputs can be Hi-(+5 volts) at a time. When the clock pulse tells the flip-flop to switch it does so. Like a light switch, once the flip-flop is switched it stays switched (it remembers that it has been switched). It has to be switched again to make change conditions.

As shown in figure 31-3, we can put a few gates together to make a circuit to add binary numbers. A computer or pocket calculator must be able to do addition. Because computers and pocket calculators add with binary numbers they must have a binary adder circuit.

If we are building a calculator we don't want our answers in binary numbers, so we can take a few more

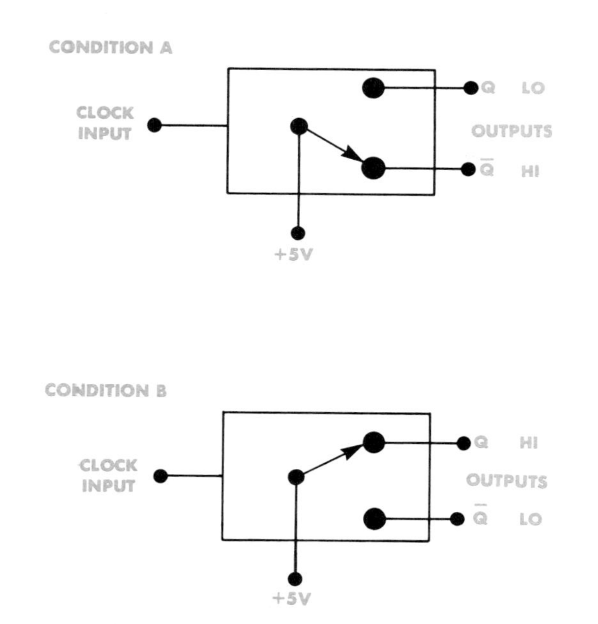

Figure 31-4: This is a model of a Flip-flop.

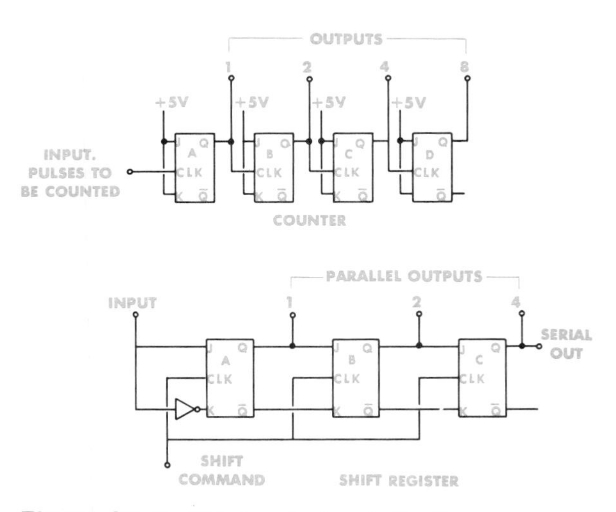

Figure 31-5: We can connect flip-flops to count in binary or we can connect them as a shift register. The shift register can shift the binary number stored in the flip-flops one position to the right on command.

gates and make a decoder that will convert binary numbers into decimal numbers.

If we have some gates connected as flip-flops we can make a counter. Figure 31-5 shows how to hook up the flip-flops. There are several IC counters we can buy with the flip-flops already hooked up. If we want to be able to read the count on a display like the one on a calculator, we can hook up a decoder. There are several kinds of decoders we can buy in an integrated circuit package.

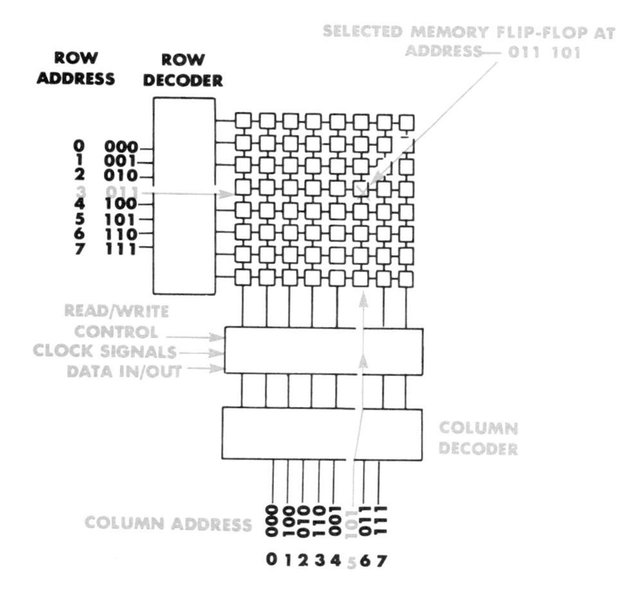

Figure 31-6: We can put a lot of flip-flops together with a couple of decoders and a few gates to make a computer memory. This kind of memory is called a Random Access Memory or RAM.

We can also hook up flip-flops to make a shift register. The shift register is a small memory unit that lets us shift 1's and zeros around. They come in IC packages with from 4 to over 1000 flip-flops.

If we take a couple of decoders and from 64 to a thousand (or more) flip-flops and a few gates, we can make a computer memory.

A memory made of 64 flip-flops is shown in figure 31-6. Memories come in IC packages with from 64 to several thousand flip-flops. The decoders are usually in the memory IC package too.

If we just keep putting these building blocks together we can end up with a pocket calculator or a full-sized computer.

SELF CHECK

1. Name the two non-inverting gates.
2. Name the three inverting gates.
3. True or False: The number system used in digital circuits is the decimal system.
4. Write the number 9 in binary numbers.

BUILDING PROJECTS WITHOUT PRINTED CIRCUIT BOARDS

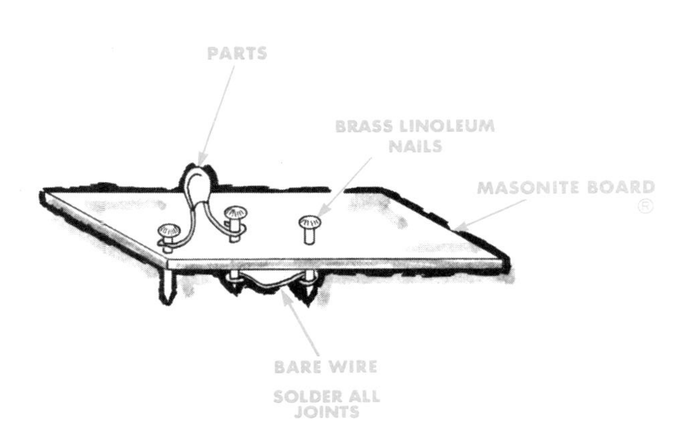

1. **Materials**
   1/8 inch thick pressed wood (masonite)
   Brass linoleum nails
   #24 bare copper wire
2. **How to do it**
   a. Lay out the positions for pads on the masonite board just as you would if you were laying out a copper clad board.
   b. Instead of sticking a paper spot on each punch-mark, drive a brass linoleum nail through the board. The nail should stick out of both sides of the masonite board.
   c. Instead of placing tape between dots you will connect bare wire between nails. Follow the tape pattern. Solder all nail-wire joints.
   d. Turn the board over and connect the parts as shown in the drawing.

WORKING WITH SHIELDED CABLE

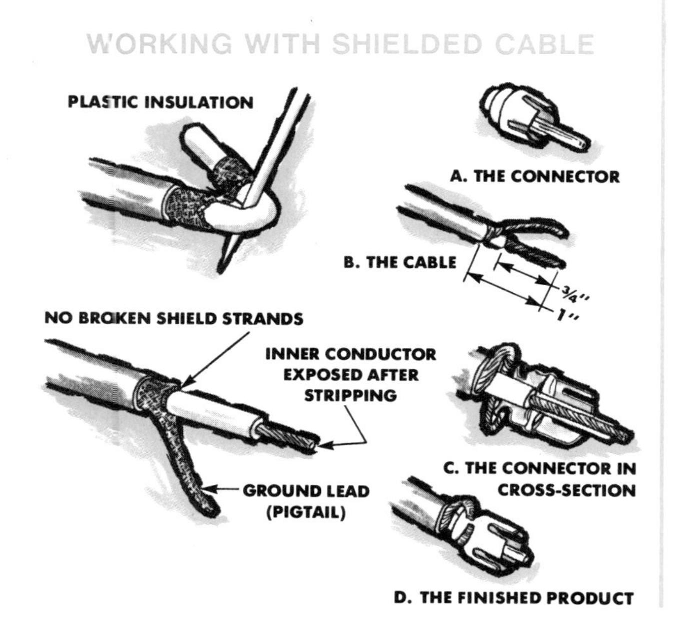

WIRING PRINTED CIRCUIT BOARDS

AN EASY WAY TO MAKE PLUGS AND SOCKETS

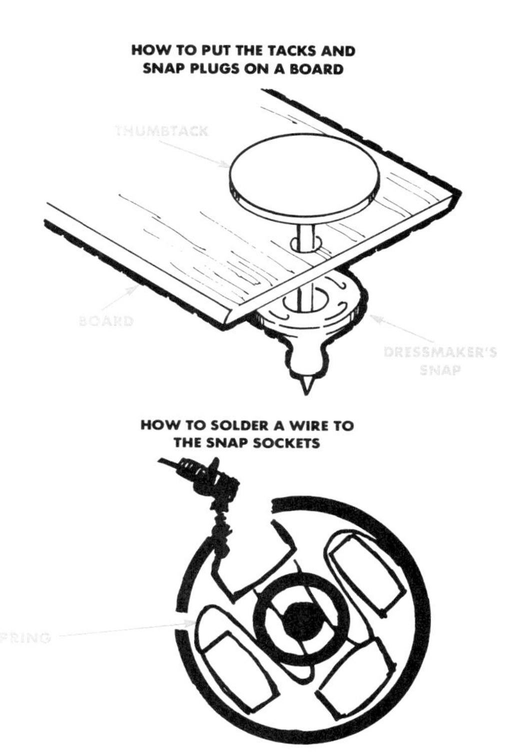

Plugs and sockets can be made from dressmaker's snaps and thumbtacks. Dressmaker's snaps are cheap. You can find them in the sewing section of many stores. It is a tight fit, but with a little pressure, a thumbtack will slip into the socket half of a snap. The circuit board should be squeezed between the tack head and the snap when you are through. Use a tiny bit of solder to join the snap and tack point. The solder will keep the snap from working loose.

The snap half has a tiny spring that goes all the way around the rim. The two free ends of the spring point to one of the holes in the snap. Snip through the edge of this hole with a pair of diagonal cutting pliers as shown in the drawing. Bend out a tab to solder wire to. If you cut through any other hole you will cut the little spring.

It is important to bend out the little tab to solder to. If you don't it is very hard to keep from soldering the little spring to the body of the snap. You may ruin a snap or two before you get the hang of it, but keep trying. Soldering miniature things takes a little practice.

This type of connector can be used for many of the projects in this book.

CONNECTING RELAYS TO CONTROL APPLIANCES

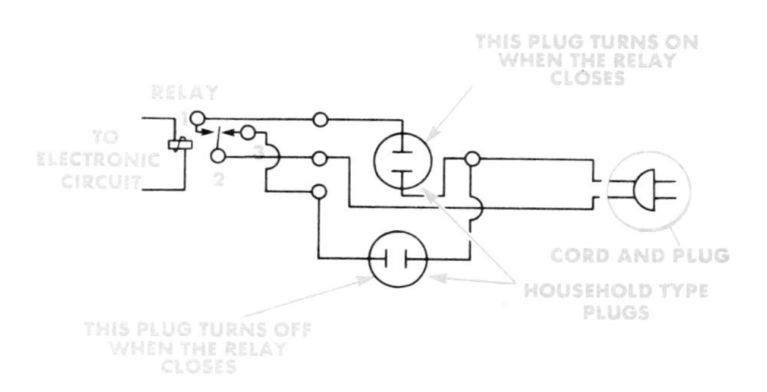

PLANT MOISTURE MINDER

Parts List

1 General purpose transistor
1 9 Volt battery and battery clip
1 Circuit board
1 Light emitting diode (LED)
1 3900 Ohm resistor (3.9K)
1 Resistor selected experimentally
1 Plastic salt shaker for case

SCHEMATIC DIAGRAM

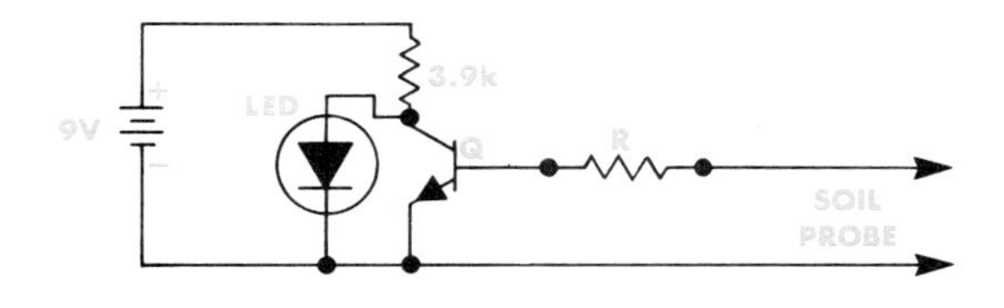

CIRCUIT BOARD LAYOUT

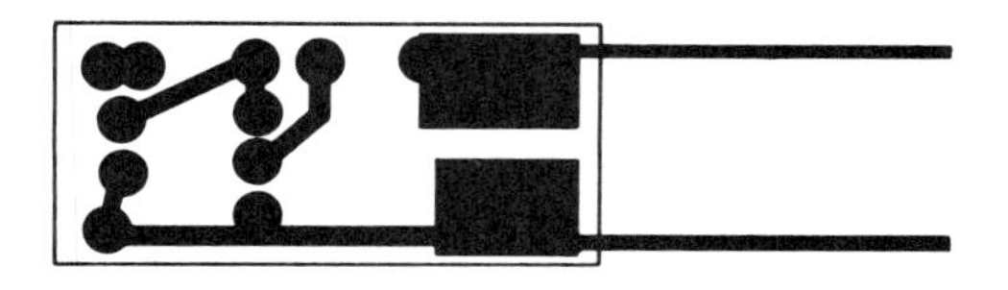

WIRING DIAGRAM

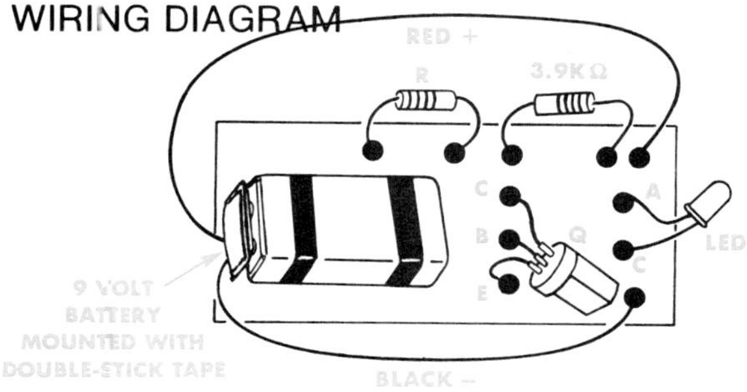

Adjustment

The value of R must be selected for your particular kind of soil. To find the right value, solder short wires where the resistor should go. Experiment with different resistor values until you find one that turns the LED out in moist soil, but turns it on in dry soil. Once you have the right value you can use that value in as many moisture minders as you want to build. Insert this gadget into the soil, and the LED will light when your plants need water.

HI-LO LIGHT DIMMER

This is a simple but useful dimmer. It can be used for lamps with up to 100 watt light bulbs. It cannot be used with fluorescent lamps.

WIRING DIAGRAM

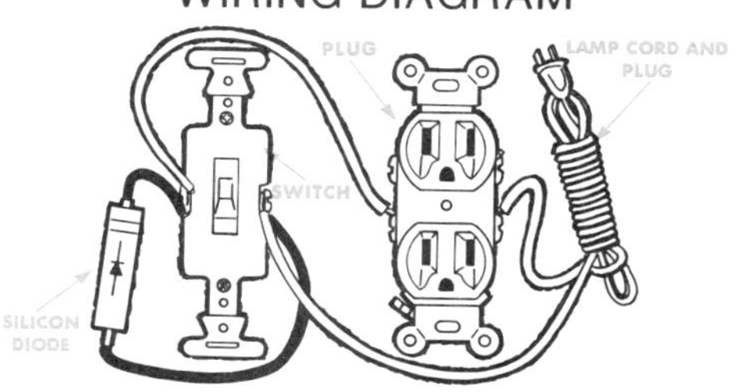

Parts List

1 House type outlet plug
1 House type wall switch
1 Lamp cord with plug
1 Silicon diode
1 Fancy switch cover
1 Fancy plug cover
1 4-inch by 5-inch picture frame or home made box

LIGHT OPERATED RELAYS

LIGHT OPERATED RELAY (BATTERY POWERED)

Parts List

1 Cadmium sulphide photoelectric cell
1 Relay
1 SPST (single pole single throw) slide switch
1 9 Volt battery and battery clip

WIRING DIAGRAM

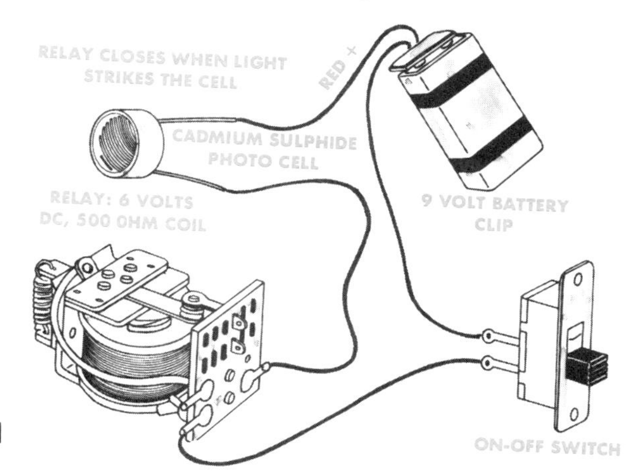

LIGHT OPERATED RELAY (PLUG-IN)

Parts List

1 Cadmium sulphide photoelectric cell
1 Relay
1 SPST (single pole single throw) switch
1 Lamp cord and plug
1 6.3 Volt transformer
1 Silicon diode
1 100 Microfarad, 25 volt capacitor

WIRING DIAGRAM

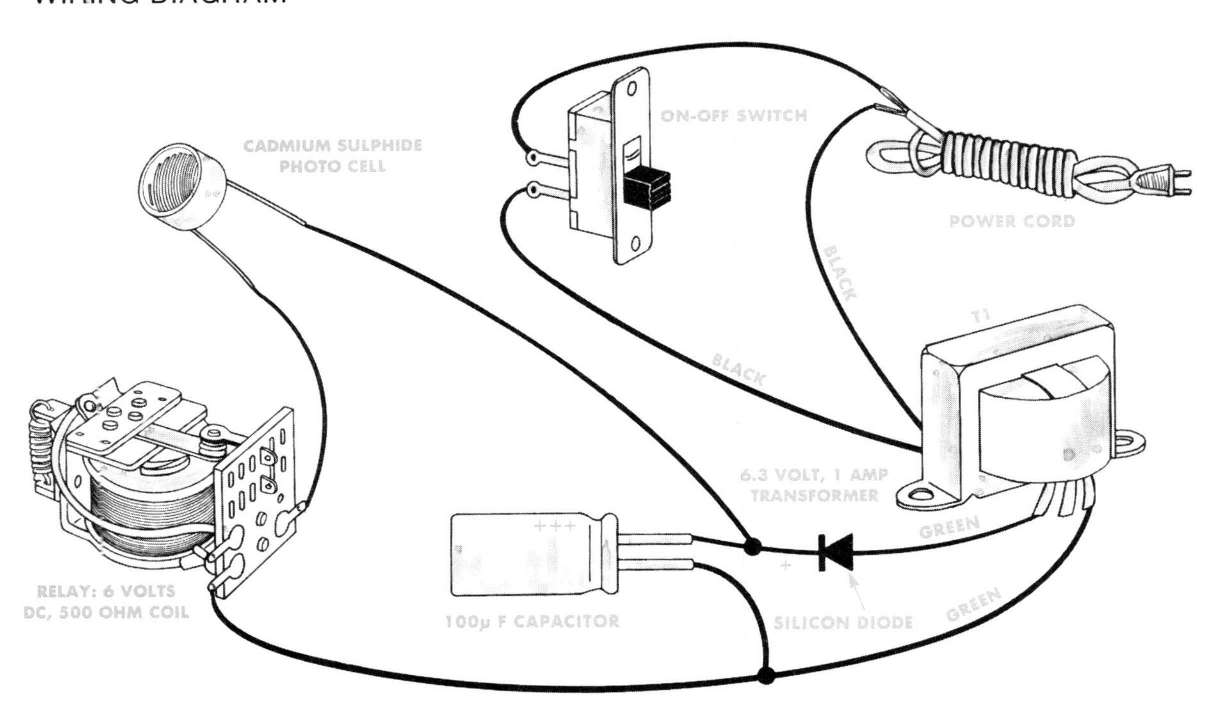

LIGHT OPERATED RELAY WITH A MEMORY

WIRING DIAGRAM

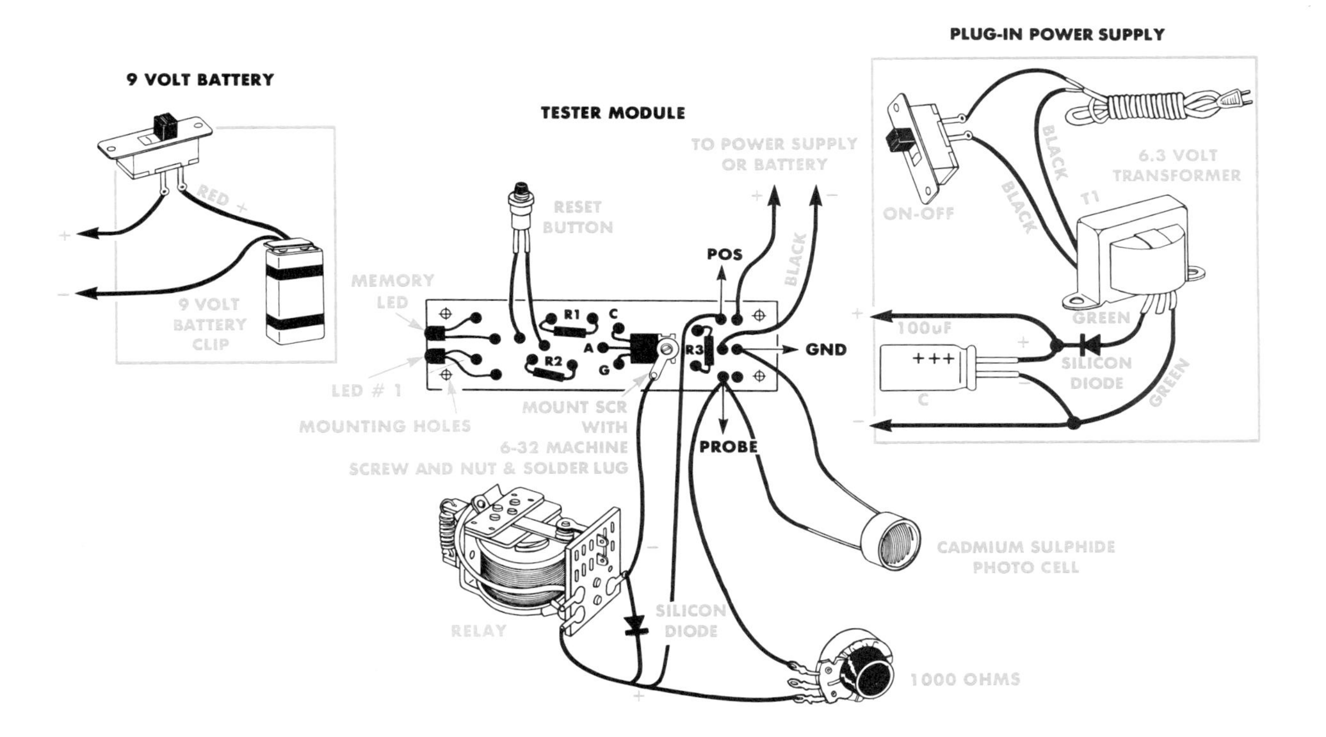

This light operated relay is very flexible. It can be either battery or plug-in powered. When it is battery powered, breaking a light beam will cause the photocell to close the relay. The relay will then stay closed until the reset button is pushed. This kind of operation is necessary in alarm systems. The unit will also lock-up when the plug-in power supply is used. If you don't want the lock-up action, remove the 100 microfarad capacitor from the plug-in power supply and connect it across the relay coil terminals. The relay will then open as soon as the light hits the photocell again.

Parts List

1 Tester module
1 Cadmium sulphide photoelectric cell
1 Relay
1 Silicon diode
1 Trimmer potentiometer (1000 Ohms)
(See the wiring diagram for power supply parts)
1 Case for the unit. This example is built in a case made of plastic pipe similar to that used for the tester.

STEREO AMPLIFIER

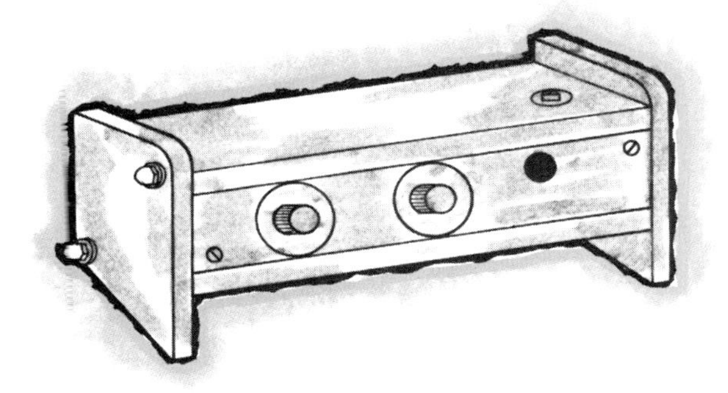

Amplifier and Power Supply Module

This module is the heart of the following projects:

1. Stereo amplifier system
2. Home or car intercom system
3. Talk on a light beam "CB" transmitter/receiver
4. Talk through the ground transmitter/receiver
5. Electronic siren
6. Emergency road flasher
7. Experimenter's power supply

The circuit board is divided into three sections. Sections A and B are two identical audio amplifiers. Section C is the power supply. Each section is a complete circuit. The board is laid out so that it can be cut apart along the dotted lines after it is etched. The stereo amplifier system uses the whole board. Most of the other projects use only one or two of the sections.

The power supply module can provide any of a number of different voltages. The regulator integrated circuit is available in several voltage ratings.

Parts List (Amplifier)

For one section:

1 General purpose transistor (Q1)
1 Power transistor (Q2)
1 6.8 microfarad tantalum capacitor (C2)
1 22 microfarad tantalum capacitor (C1)
1 Volume control potentiometer (R1) 100K Ohms audio taper

CIRCUIT BOARD LAYOUT

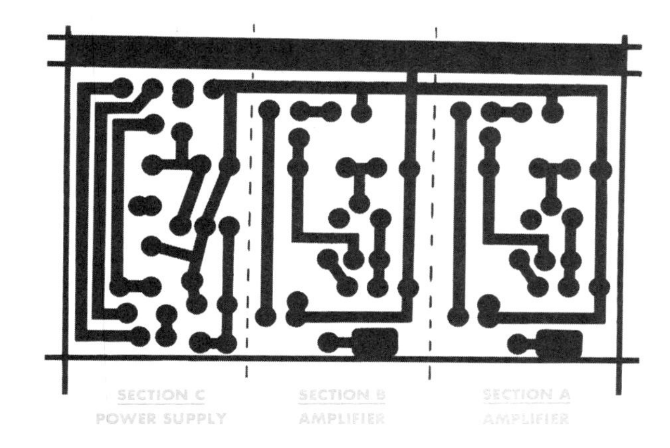

1 Trimmer potentiometer (R3) 100K Ohms
1 39K Ohm resistor (R2)
1 RCA type phono jack (P1)
1 Binding post for GND connection (G). Used for section A only
1 Screw type terminal strip for speaker connections. One is used for both amplifier sections.
1 Solder lug (part of power transistor mounting hardware)
1 Circuit board (section A or section B)

Parts List (Power Supply)

1 Circuit board
1 6.3 Volt, 1 amp transformer (T1)
1 Power cord and plug
4 Silicon diodes
1 Voltage regulator integrated circuit LM340T-5 (VR)
1 Electrolytic capacitor, 500 microfarad, 25 volts (C3)
1 Light emitting diode (LED)
1 SPST slide switch
1 Resistor, 150 Ohms

WIRING DIAGRAM

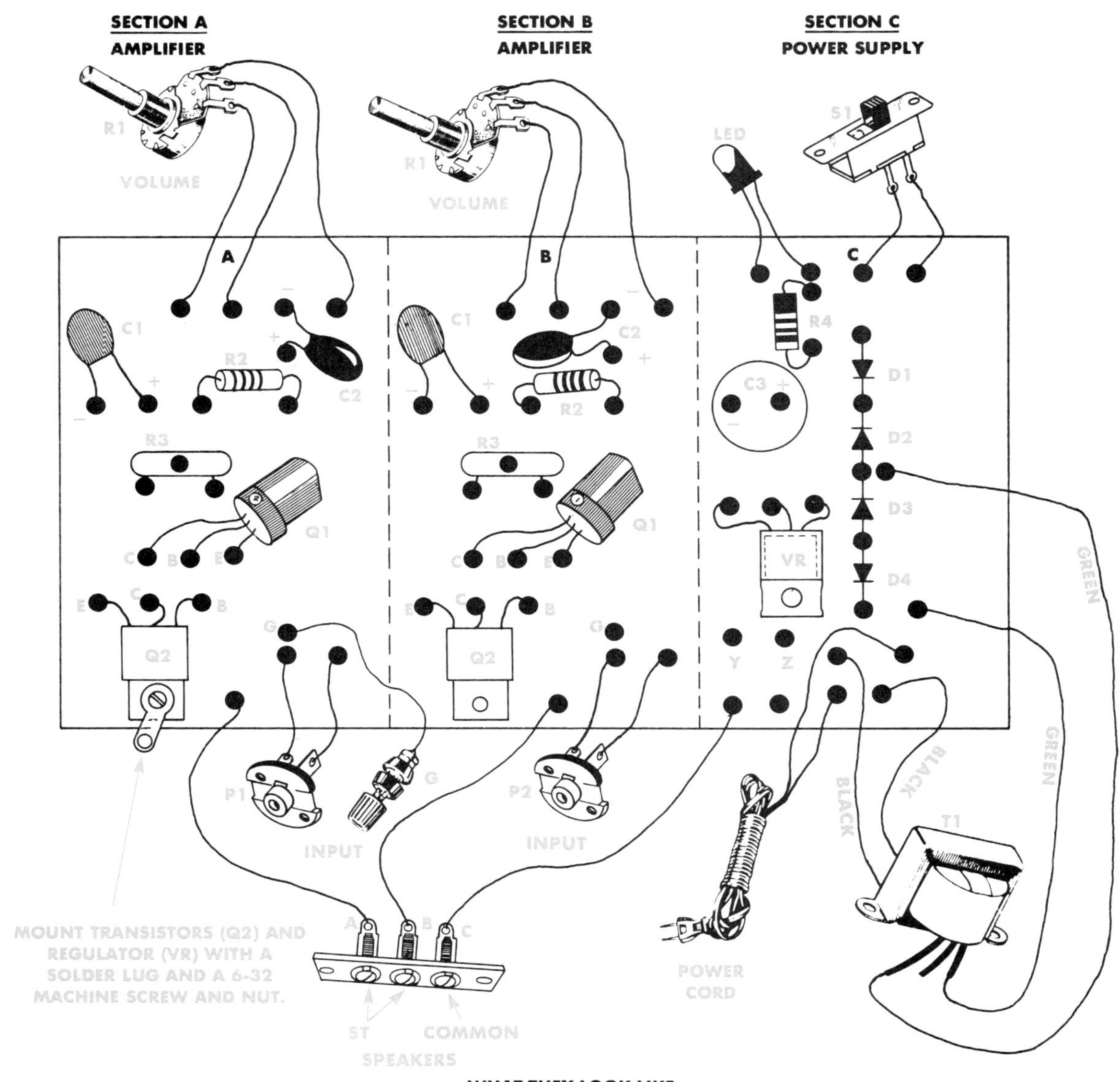

WHAT THEY LOOK LIKE

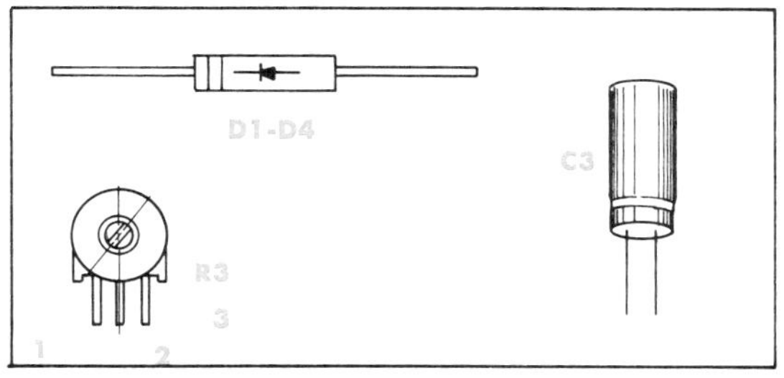

SCHEMATIC DIAGRAM

AMPLIFIER
(BOTH SECTIONS ARE THE SAME)

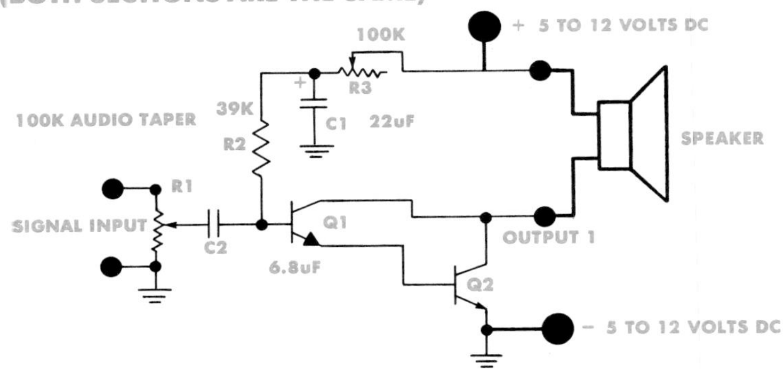

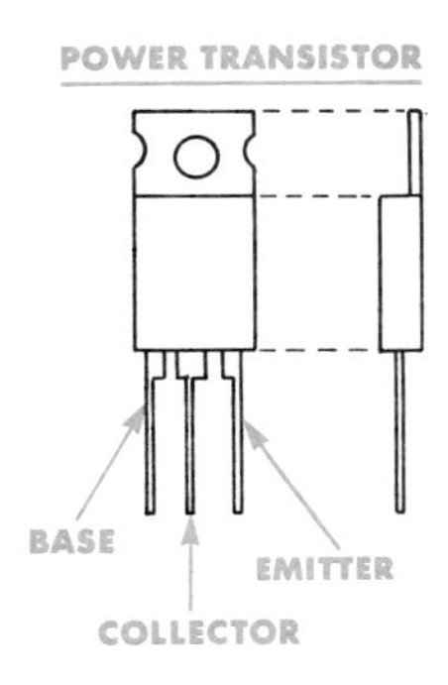

POWER SUPPLY

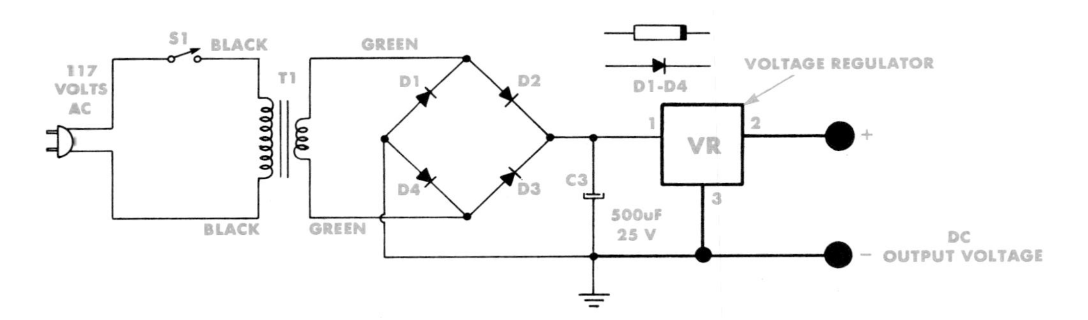

BASING DIAGRAM

NOTE: FOR BATTERY OPERATION USE 4 SIZE D CELLS OR A 6 VOLT LANTERN BATTBATTERY. THE AMPLIFIERS WORK FINE WITH THE ELECTRICAL SYSTEM IN A CAR.

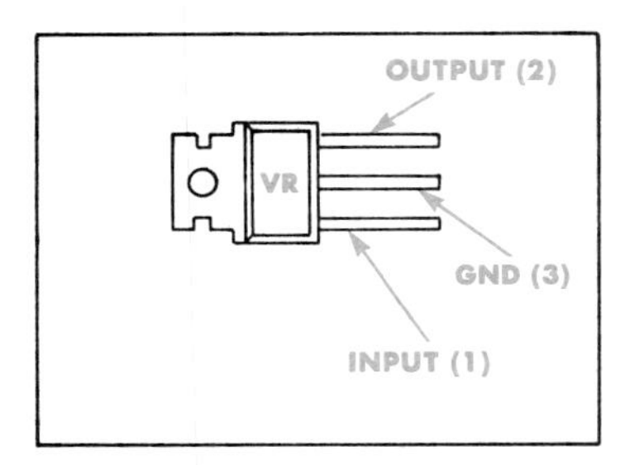

Amplifier Adjustments

1. Connect the speaker to channel A output. Connect a tuner or microphone to the input of channel A. If you use the microphone, place a transistor radio in front of the microphone. The radio will provide sound to adjust with.
2. Adjust the amplifier volume control to about the middle of its range. Now adjust R3 on the circuit board for maximum volume. Repeat the procedure for channel B.

MAKING A CABINET

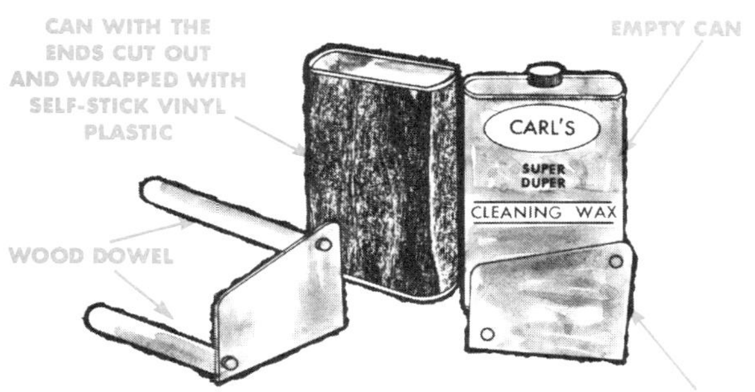

Instructions

1. Wash empty can with dishwashing detergent.
2. Remove the ends of the can with a can opener.
3. Wash the can with detergent, inside and outside. Dry thoroughly.
4. Cut self adhesive plastic from roll and wrap around can. (You can get the plastic in any supermarket)
5. Cut end pieces out of masonite and stain or paint them. The end pieces are held to the can with wood dowel spacers and wood screws.

HOOKING UP THE SPEAKERS WHEN THE BUILT-IN POWER SUPPLY IS USED

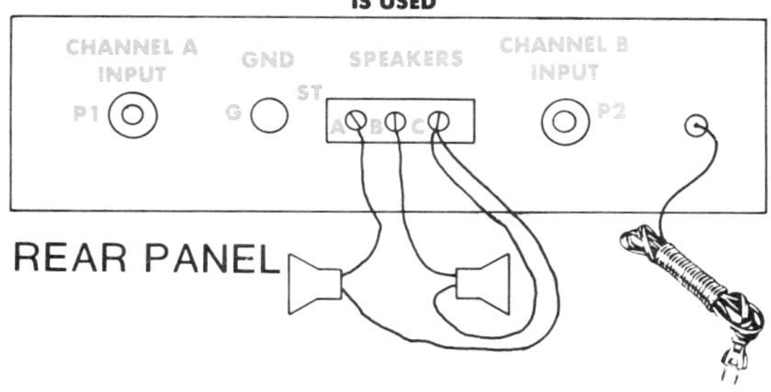

HOOKING UP THE SPEAKERS WHEN YOU USE A BATTERY

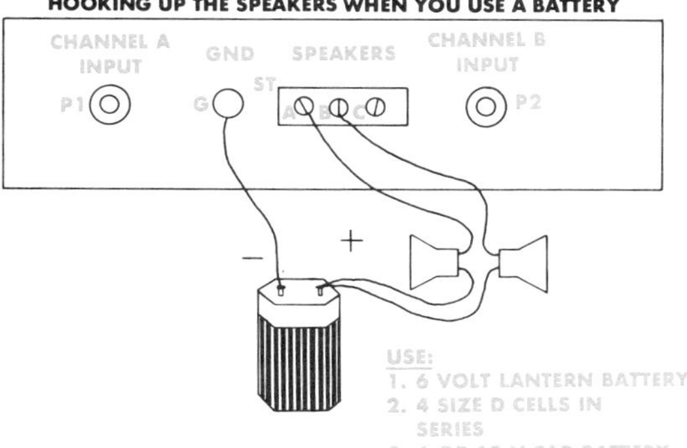

AM TUNER

This project is an easy one. Buy an inexpensive transistor pocket radio. The sound quality is not bad at all when the radio's cheap amplifier and speaker are replaced by an amplifier and speaker with better quality.

Making the Connections

1. Cut the wires that go to the speaker and earphone plug.
2. Find the volume control. Solder a short piece of wire to the center lug. There are three lugs on the control. Solder another short piece of wire to the negative battery wire. Connect the wires to a phono connector as shown in the figure.
3. Remove the speaker and mount the tuner in a case. If you use the kind of case shown for the stereo amp make sure the antenna coil is away from metal and near the masonite end.

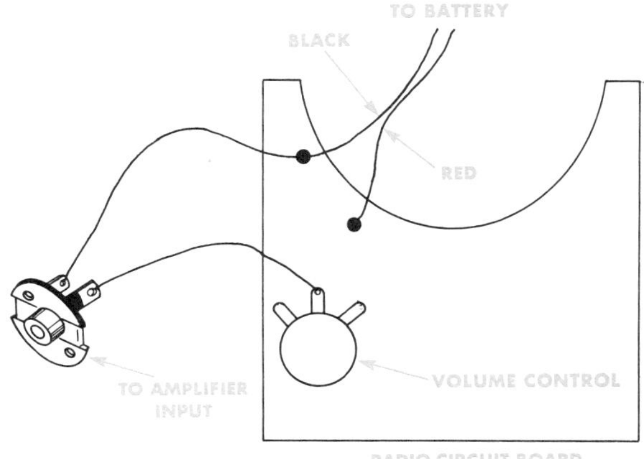

MICROPHONE/SPEAKER MODULE

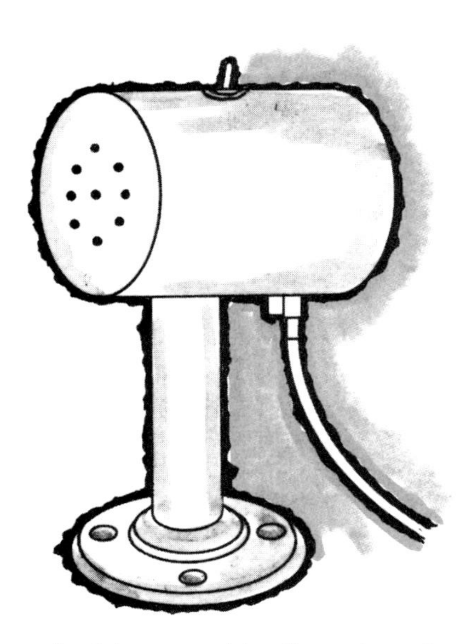

This project is a combination microphone and speaker. It can be used with the stereo amplifier module to make the following projects:

1. Intercom system
2. Talk on a light beam "CB"transmitter/ receiver
3. Talk through the ground "CB" transmitter and receiver

It can also be used as a general purpose microphone or test speaker.

SCHEMATIC DIAGRAM

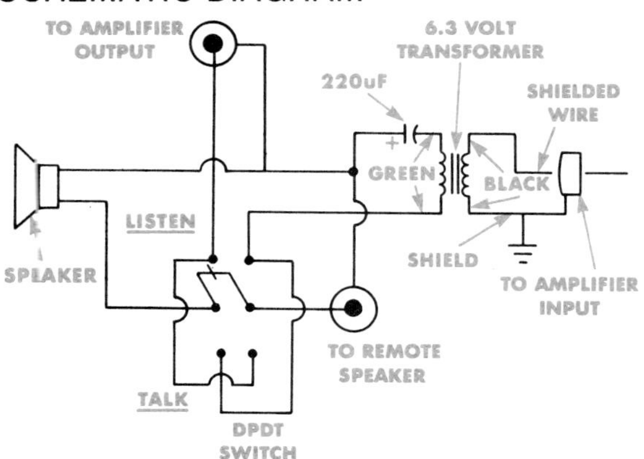

WIRING DIAGRAM

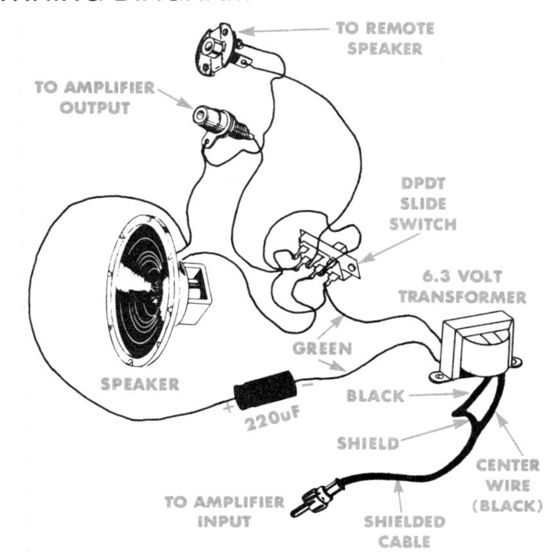

Construction Hints

Instead of a circuit board, use a small piece of masonite and linoleum nails (brass). A large plastic drinking cup, some plastic pipe, and a cast iron pipe flange (base) were used in the example for a case. The cup must have thick walls and the open end must have a cover. A plastic bowl might be easier to fit the parts in. The cup is very hard to get the parts into.

SWITCH AND PLUG PANEL

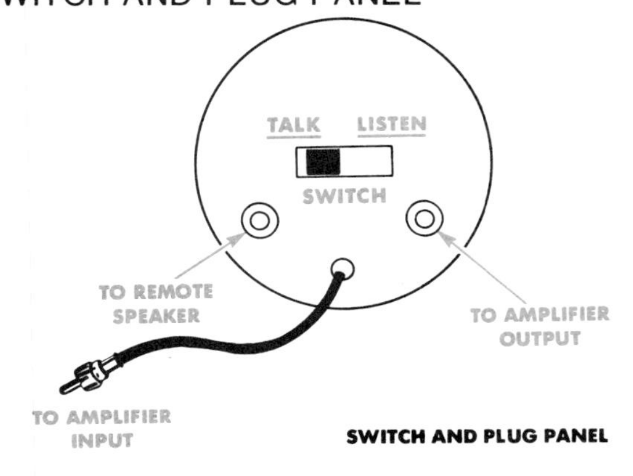

SWITCH AND PLUG PANEL

INTERCOM SYSTEM

This project uses one channel of the stereo amplifier, the microphone/speaker module, and an extra speaker.

Parts List

1 Sections B and C—stereo amplifier module
1 Microphone/speaker module
1 Extra (remote) speaker and box to put it in
25 to 100 feet of intercom or speaker wire

WIRING DIAGRAM

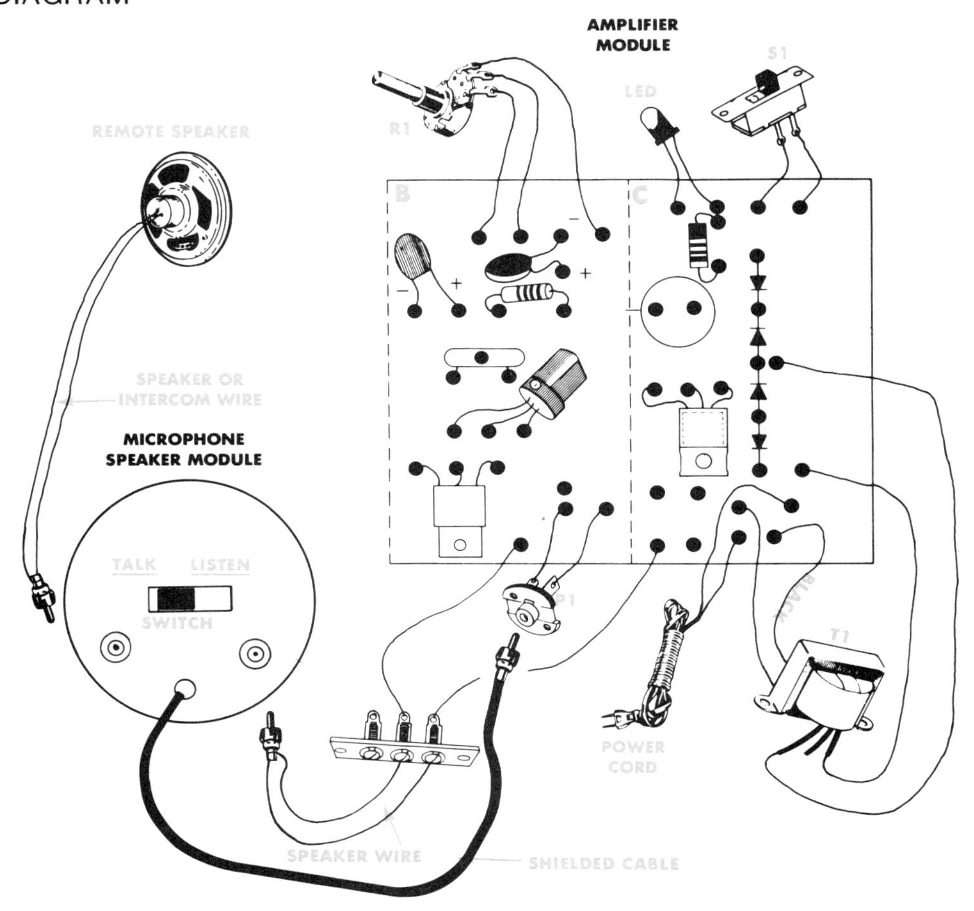

TALKING OVER A LIGHT BEAM "CB" TRANSMITTER/RECEIVER

MCUNTING THE PHOTO TRANSISTOR

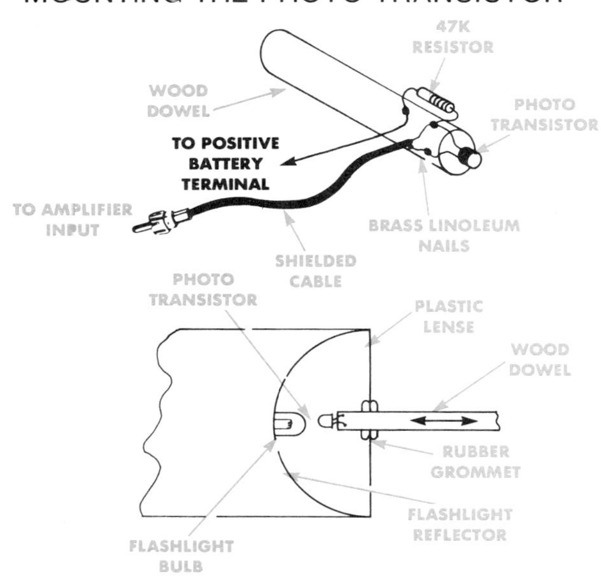

This project allows you to talk back and forth over a light beam. You can get a range of a city block or more at night. Daylight operation is possible up to about 200 feet. The system requires two stations. Each station uses both channels on the stereo amplifier board (A and B). The power supply section is not used because the units are battery powered. Each station also uses a speaker/microphone module. A lantern type flashlight is used for both transmitting and receiving. The flashlight bulb produces the transmitting beam. A photo transistor receives the beam coming from the other station. The flashlight reflector concentrates the beam for transmitting and gathers the light for the photocell to receive. The voice quality is very good.

Adjustments

1. Place the two stations about 10 feet apart.
2. Tape the press-to-transmit button down on one station. Switch the unit to "TALK". Place a transistor radio near the microphone to talk for you while you make the adjustments.
3. Set the other station to "LISTEN". Move the receiving station around in the beam until you get the loudest volume.
4. Move the wood photo transistor support rod back and forth until you get the loudest volume.
5. Now go back to the transmitter and adjust R3 on the A section of the circuit board for the clearest and loudest signal from the receiver.
6. Repeat the procedure in the opposite direction. These adjustments should only have to be made once.
7. Now try for some distance.

NOTE: For greater distance you can use two lantern batteries in series to get 12 Volts. You should get much better range, but the flashlight bulb will not last as long. You may have to adjust R3 again if you try 12 volt operation.

Parts List (For One Station)

1 Stereo amplifier module, sections A and B. The power supply section (C) is not used.
1 Microphone / speaker module
1 Lantern type flashlight
1 NORMALLY OPEN push button (push to transmit)
1 SPST slide switch (ON-OFF)
1 Photo transistor
1 47,000 (47K) Ohm resistor
1 Wood Dowel; 6 inches long
Some brass linoleum tacks
1 Rubber grommet to fit around the dowel
2 RCA type phono jacks
6-8 Feet of shielded microphone cable

Parts List (For the Tripod)

6 Plastic pipe (PVC) tee fittings
1 45 degree elbow fitting
3 Pieces of plastic pipe ½ inch by 6½ inches long
2 Pieces of plastic pipe ½ inch by 2 inches long
1 Piece of plastic pipe ½ inch by 4 inches long
NOTE: all fittings are ½ inch

WIRING DIAGRAM

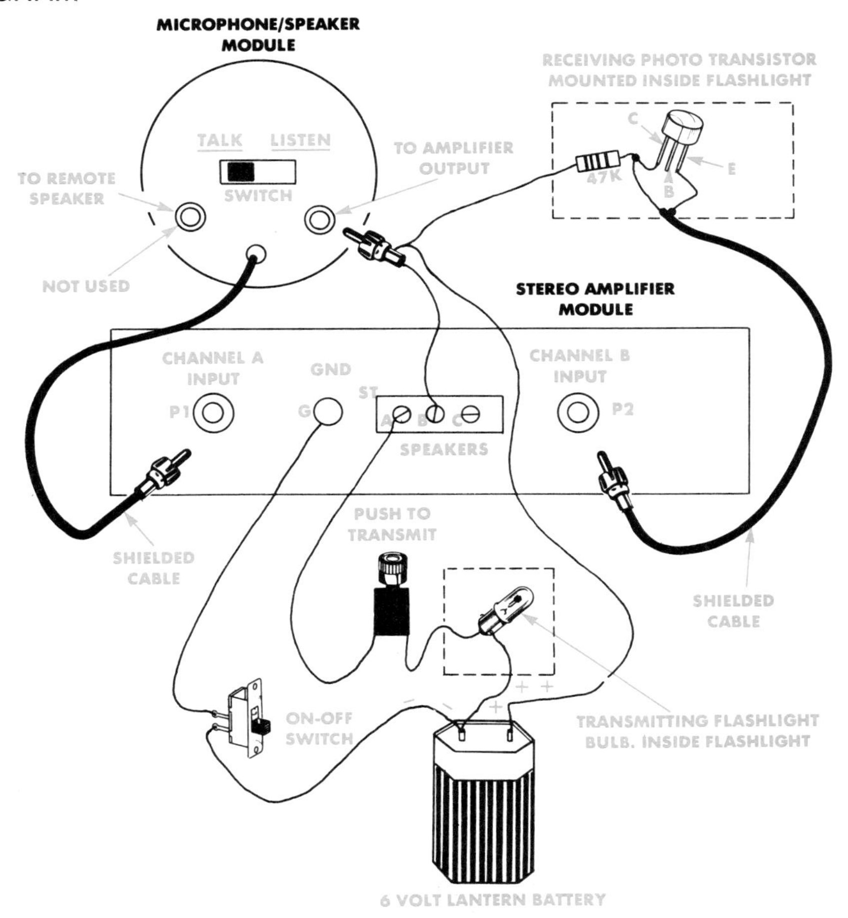

TALKING THROUGH THE GROUND TRANSMITTER/RECEIVER

This project is more of an experiment than a project. You can transmit through the ground if soil conditions are right. It can work very well, but it isn't worth building equipment until you find a good plot of grass. How well it works depends on a lot of conditions, but it is a fun experiment if you already have the amplifiers built.

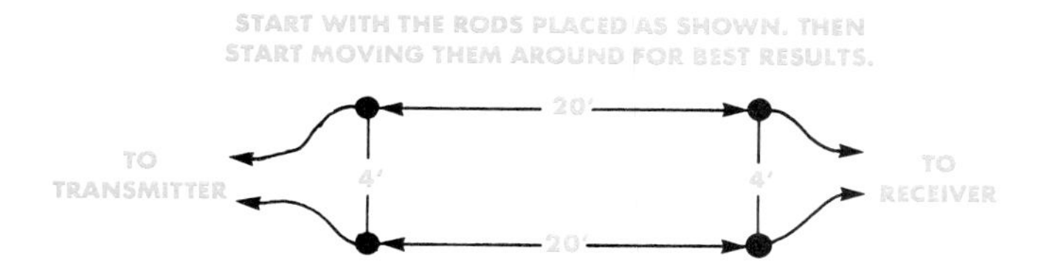

WIRING DIAGRAM (TRANSMITTER)

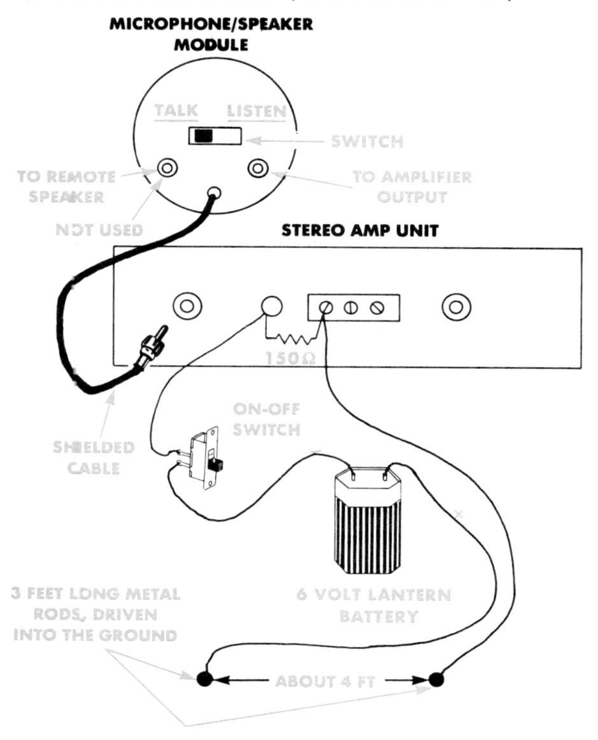

WIRING DIAGRAM (RECEIVER)

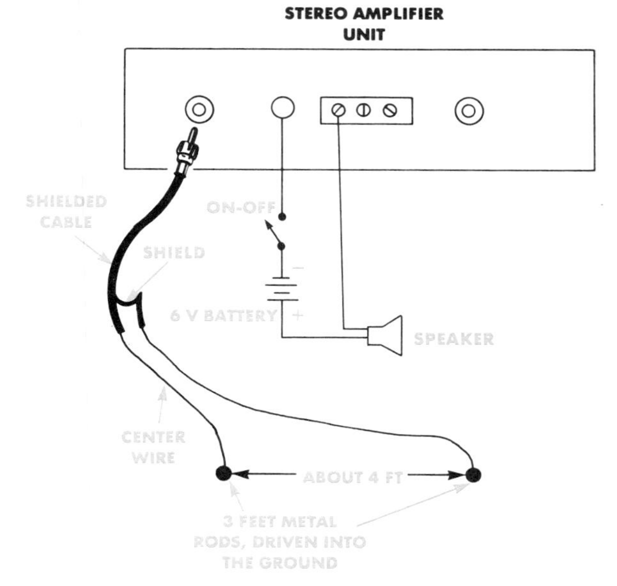

EXPERIMENTER'S POWER SUPPLY

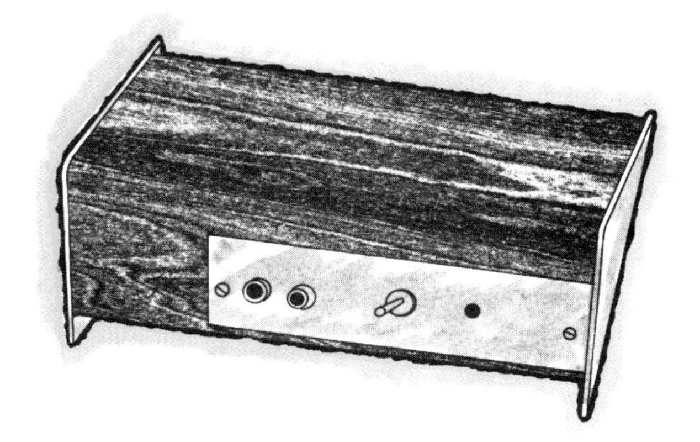

Choosing the Right Regulator and Transformer

OUTPUT VOLTAGE	REGULATOR NUMBER	TRANSFORMER VOLTAGE
5 V.	LM340T-5	6 to 12 VOLTS
6 V.	LM340T-6	10-12 V.
8 V.	LM340T-8	10-12 V.
12 V.	LM340T-12	15-24 V.
15 V.	LM340T-15	24-28 V.
18 V.	LM340T-18	24-28 V.
24 V.	LM340T-24	28-35 V.

This power supply can be used to take the place of batteries for any of the projects in this book. It can be used to power almost any other projects you might build from magazines, etc. Use the C section of the stereo amplifier board. Select the right regulator (VR) for the voltage you want. The number after the dash in the table tells you what voltage the regulator will put out. The column to the right in the table tells you what transformer voltage you need.

SCHEMATIC DIAGRAM

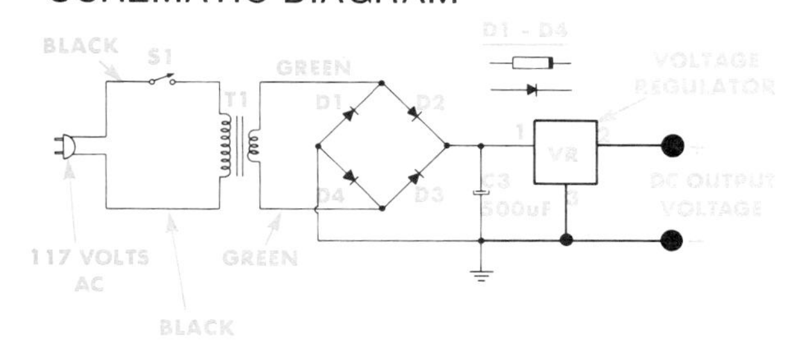

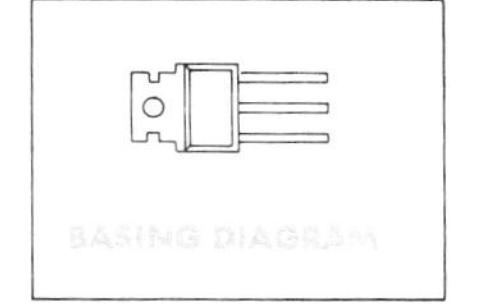

WIRING DIAGRAM

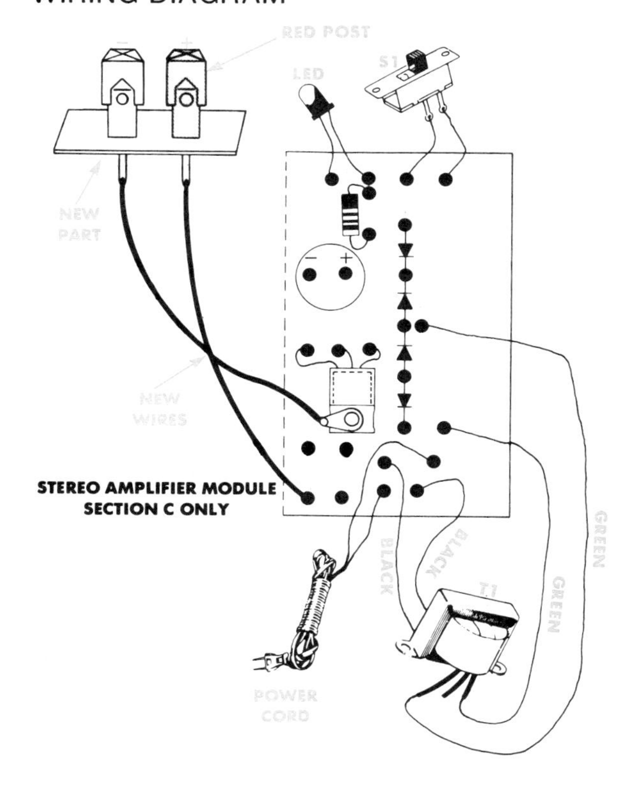

RUN DOWN CLOCK MODULE

The run down clock module is an oscillator (pulse generator). It can produce a steady tone or one that rises and falls in pitch. The pitch is adjusted with the trimmer potentiometer (R3). Capacitors C1, C2, and C3 are selected for the particular project you are building so leave them out until later.

This module can be used with the following projects:

1. The signal generator project
2. Emergency roadside flasher unit
3. An electronic siren
4. The wheel of fortune project

Parts List

1 Circuit board
1 Unijunction transistor
2 100 Ohm resistors
2 4700 Ohm (4.7K) resistors
1 Light emitting diode (LED)
1 50,000 (50K) Trimming potentiometer
1 NORMALLY OPEN push button

SCHEMATIC DIAGRAM

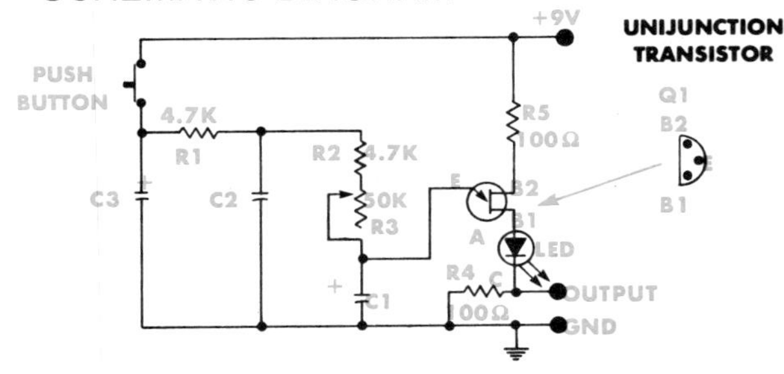

CIRCUIT BOARD LAYOUT

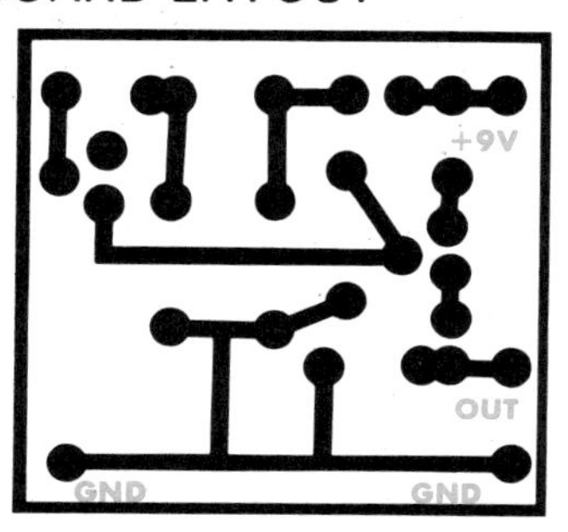

WIRING DIAGRAM

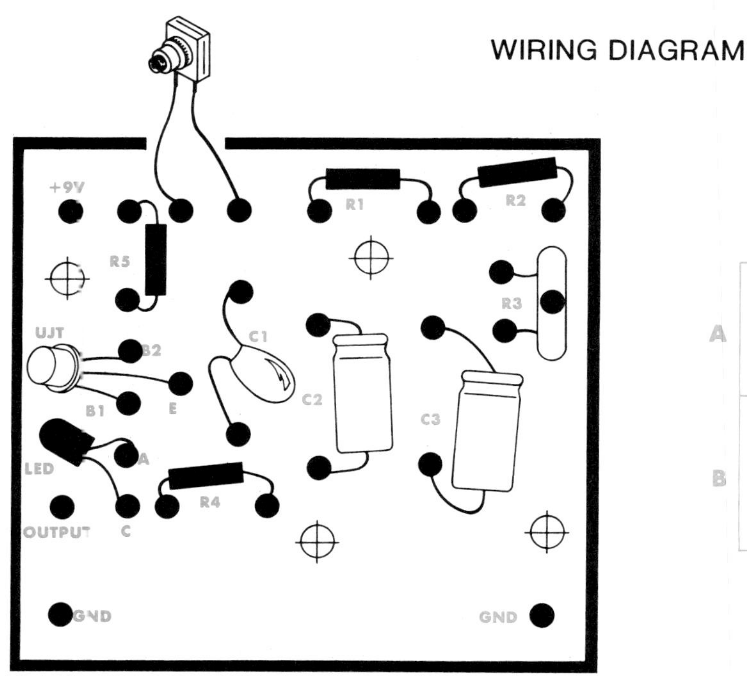

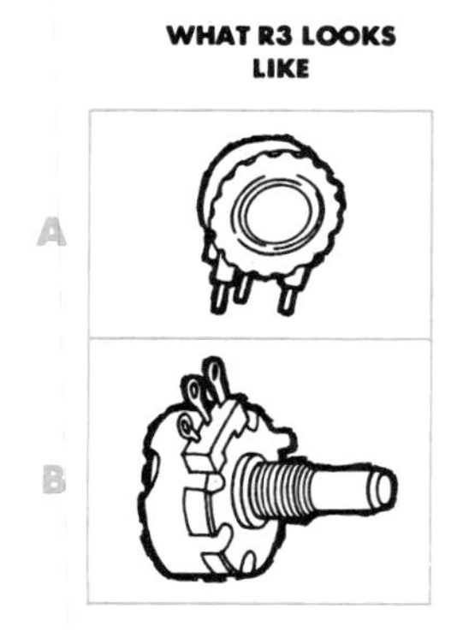

SIGNAL INJECTOR (GENERATOR)

This project is a very useful test instrument. It allows you to find a dead or weak circuit in amplifiers, radios, etc. It injects a tone that you can follow through a set. To use the injector, place the probe on the base of the transistor nearest the speaker. If you hear the tone, that transistor and the parts connected to it are okay. Move the probe back to the next transistor base. The tone should be louder. If it is weaker or gone, you have located the problem stage. As you move from transistor to transistor, away from the speaker, the volume of the tone should keep increasing. When you come to a transistor where the tone gets weaker or is lost, you have located the problem section.

The signal injector can be mounted any of the normal ways, but a three layer wood sandwich works well. Cut out the middle layer for the circuit board, battery, and push button. Shape the outside layers to look and feel good, and bolt them in place.

WIRING DIAGRAM

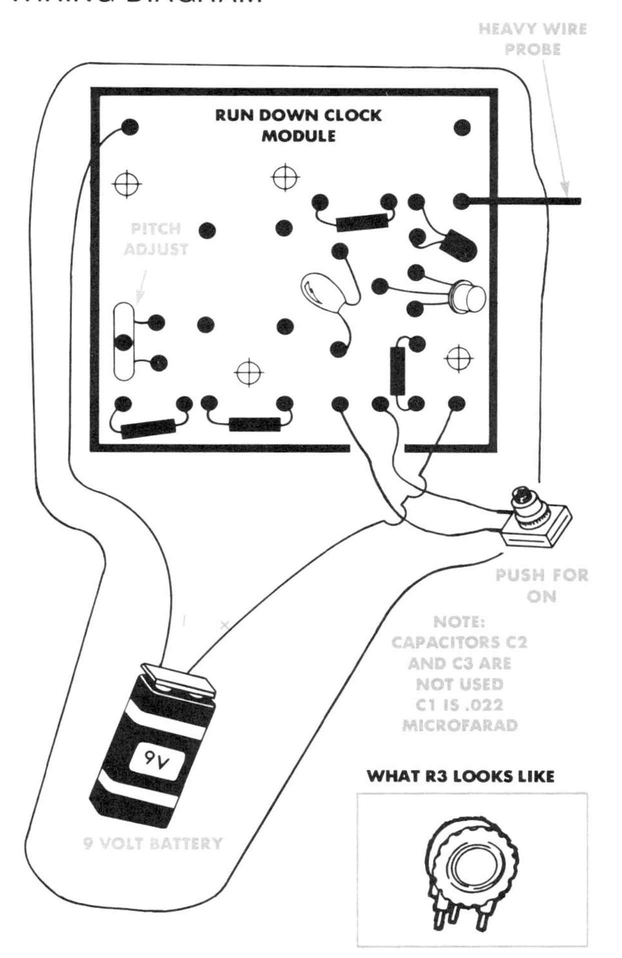

EMERGENCY ROADSIDE FLASHER

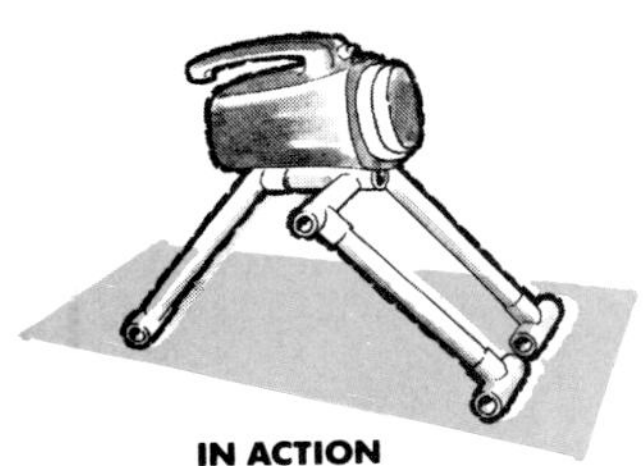

IN ACTION

FOLDED FOR STORAGE

This flasher unit produces a very bright flashing light. The flashing rate can be adjusted with R3 on the run down clock module. The battery is removed from a lantern type flashlight and the two modules are mounted inside the case. Red plastic can be fastened to the clear plastic flashlight lens. Use a tripod like the one used for the light beam transmitter. It doesn't take up much space and it gets the light up off the ground so it can be seen at a greater distance.

Parts List

1 Lantern type flashlight
1 Amplifier module (section A only)
1 Run down clock module
NOTE: Capacitors C2 and C3 are not used
Capacitor C1 is a 1 microfarad tantalum capacitor
1 Cigaret lighter plug

CONNECTING THE MODULES

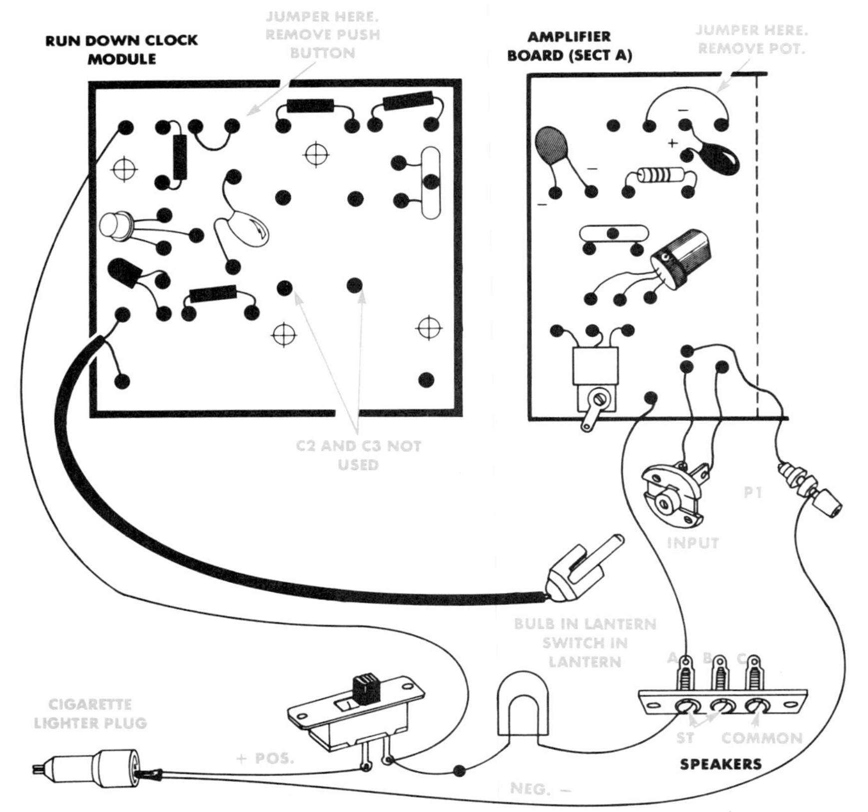

ELECTRONIC SIREN

This project is an attention-getting sound maker. It sounds very much like a police siren so don't use it in a car.

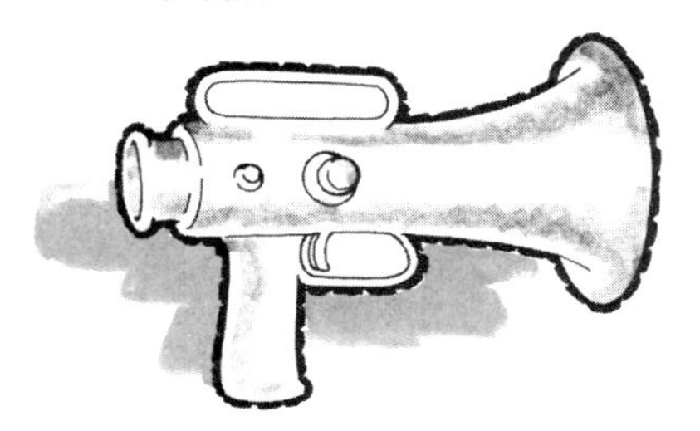

Parts list

1 Stereo amplifier module (section A only)
1 Run down clock module with the following capacitor values:
 C1=.047 microfarad (smaller capacitor gives a higher pitch)
 C2=50 microfarad (25 volts) electrolytic
 C3=100 microfarads (25 volts) electrolytic
1 Speaker
1 SPST slide switch (ON-OFF)
1 RCA type phono plug and 8 inches of shielded microphone cable
2 Lantern batteries in series for power

CONNECTING THE MODULES

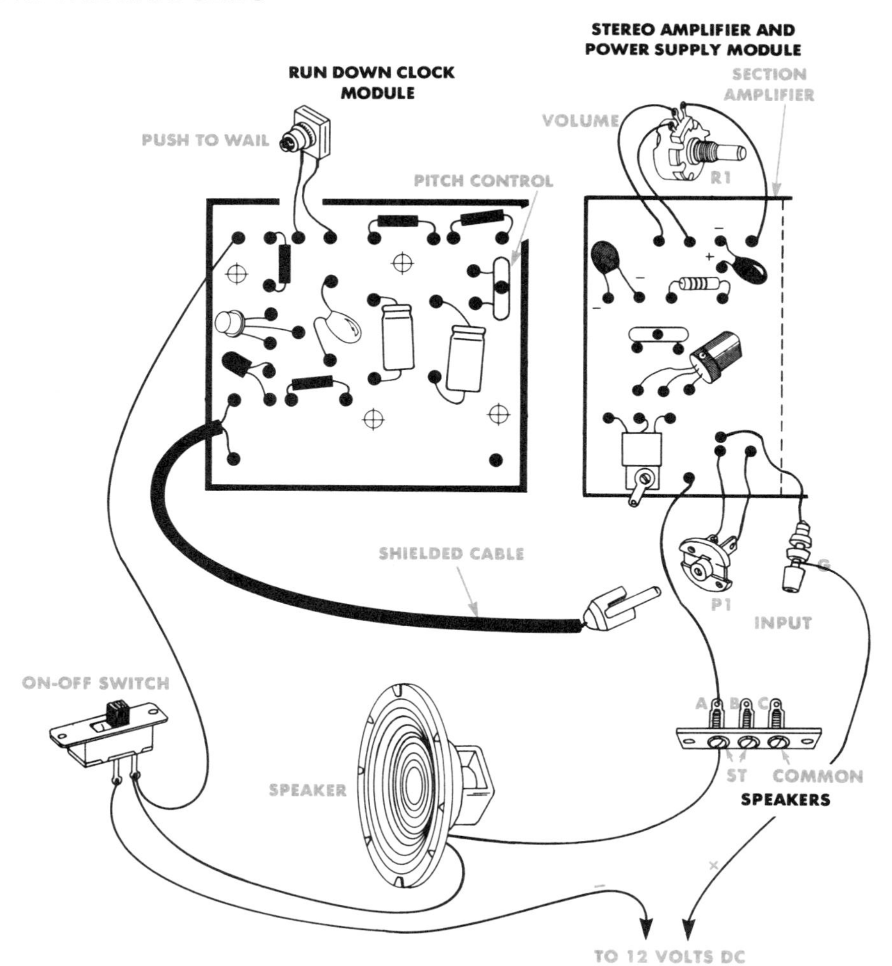

PRECISION CLOCK MODULE (TIMER)

This module uses an 8 pin DIP integrated circuit. The 555 timer integrated circuit is a precision timer or pulse generator. It is used for the following projects:

1. Space battle sound effects machine
2. A warble type alarm circuit
3. A simple kitchen or darkroom timer
4. A fancy super timer for any timing application

Parts List

1 555 timer integrated circuit
1 8 pin DIP wire wrap socket
1 .01 to .047 microfarad capacitor (C2)
1 Circuit board

NOTE: Resistors R1, R2, and R3 are put in when you build a project. The values differ from project to project. Also, leave capacitor C1 out for now.

SCHEMATIC DIAGRAM

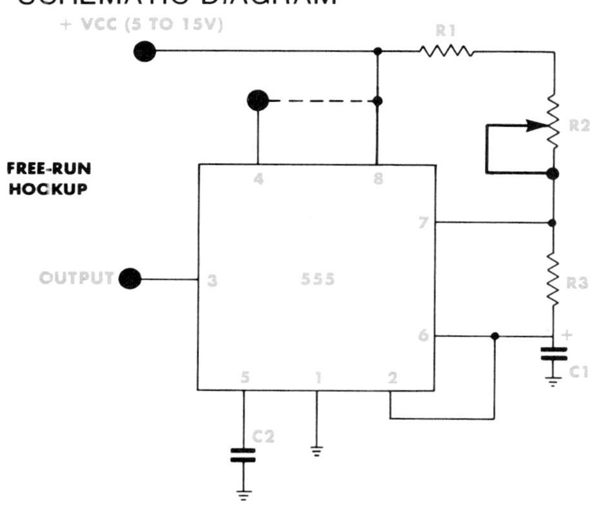

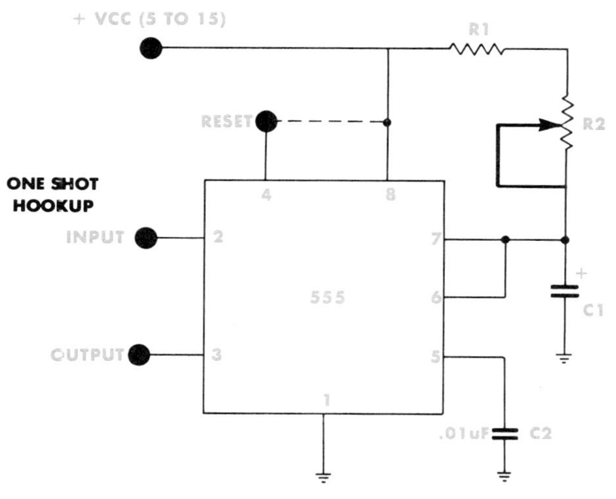

CIRCUIT BOARD LAYOUT

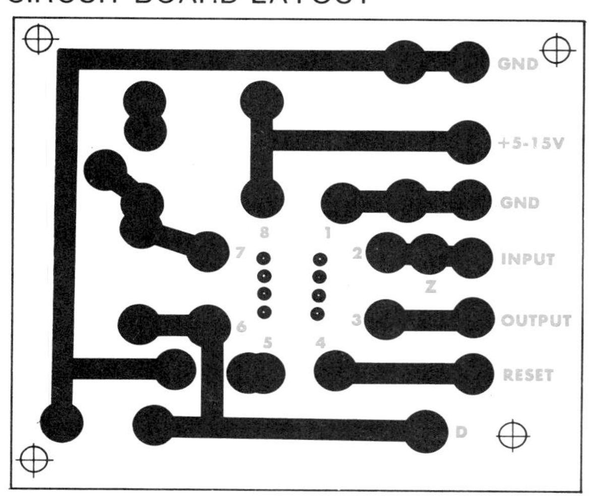

WIRING DIAGRAM

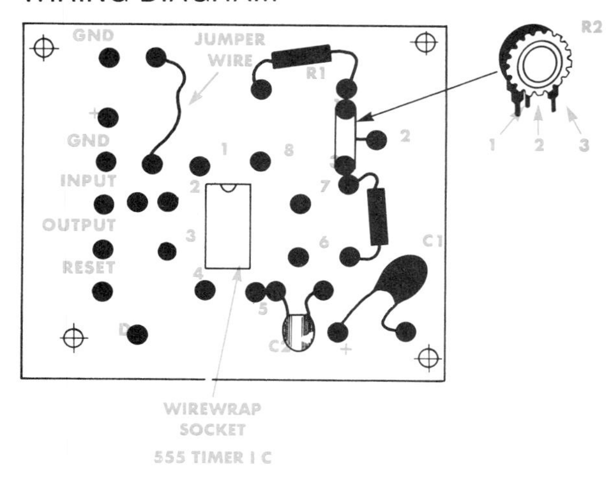

SPACE BATTLE SOUND EFFECTS MACHINE

This gadget can make an almost unending variety of weird science fiction movie sounds. Two potentiometers allow you to experiment with different sound effects.

Parts List

2 Precision clock (timer) modules
2 250,000 Ohm (250K) linear taper potentiometers (R2 and R2b)
2 1000 Ohm (1K) resistors (R3a and R1b)
2 .01 microfarad capacitors (C1a and C1b)
1 2.7 microfarad tantalum capacitor (C3)
1 NORMALLY OPEN push button (trigger)
1 SPST slide switch (ON-OFF)
1 Battery. Use a 9 volt battery if you plug into a stereo amplifier module. If you hook the speaker up directly to the unit use 4 size C cells in series (6 volts).

SCHEMATIC DIAGRAM

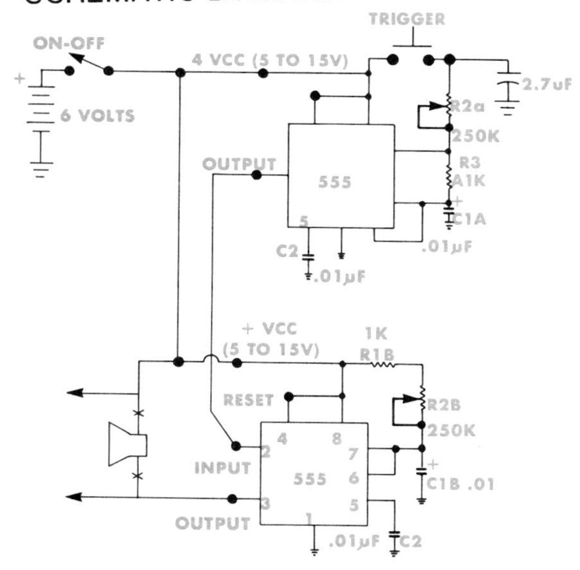

CONNECTING THE MODULES

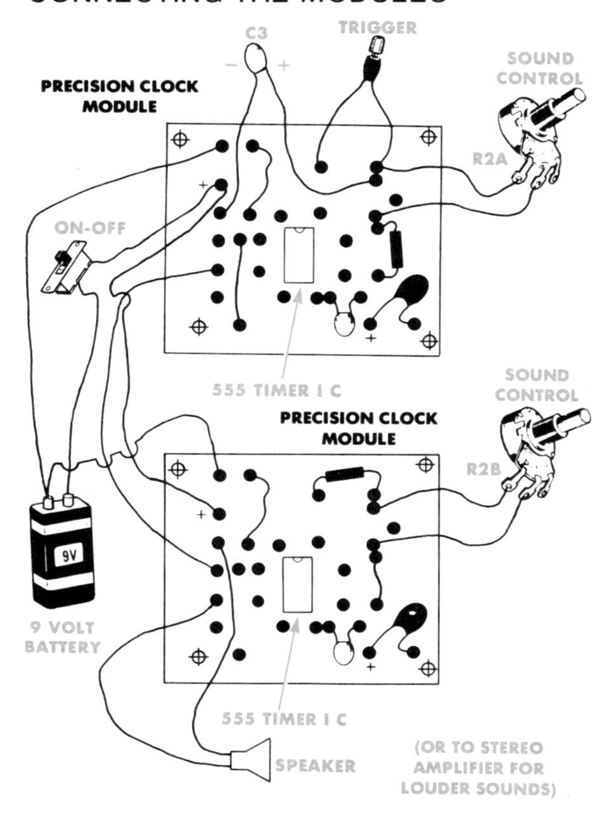

WARBLE SIREN ALARM

This alarm is a good one for burglar, car, or fire alarms. It sounds like an emergency vehicle whooping alarm. It can be used with a speaker or connected to a stereo amplifier module for more volume.

Parts List

2 Precision clock modules
1 Speaker (optional)
1 Battery (9 volt or 6 volt lantern type)
1 4700 Ohm (4.7K) resistor (R4)
1 10,000 Ohm (10K) resistor (R5)
2 1 megohm resistors (R1a and R3a)
2 33,000 Ohm (33K) resistors (R1b and R3b)
1 .1 microfarad capacitor (C1a)
1 .01 microfarad capacitor (C1b)
1 10 microfarad tantalum capacitor (C4)

NOTE: Capacitor C2 must be removed from clock module B.

SCHEMATIC DIAGRAM

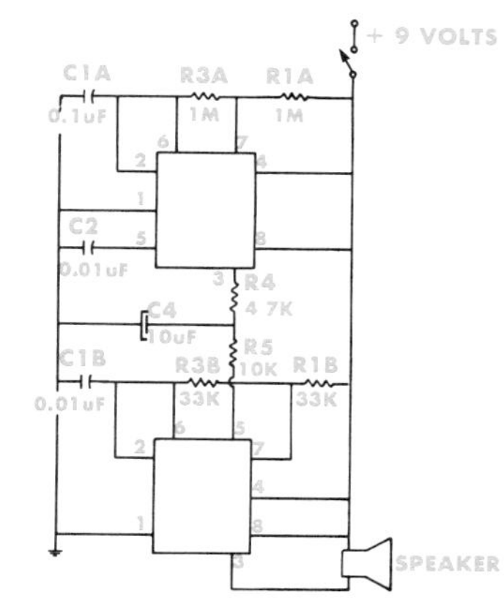

CONNECTING THE MODULES

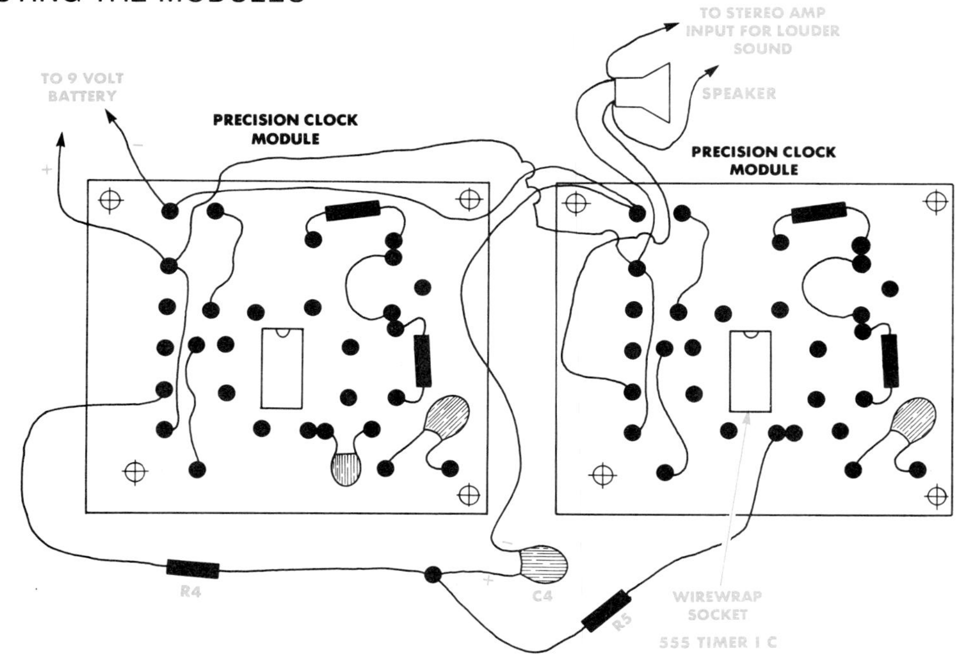

SIMPLE TIMER

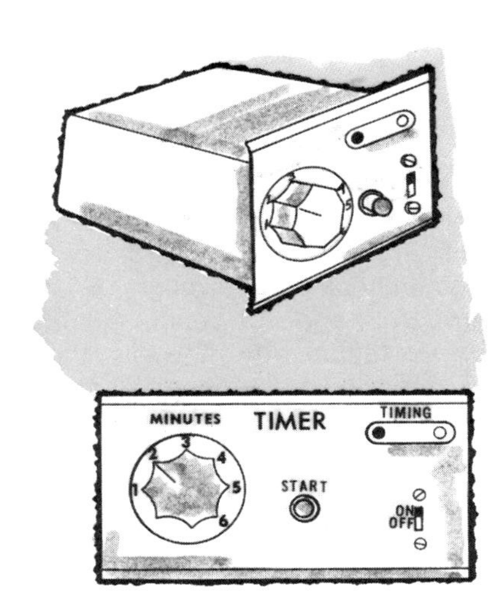

WIRING DIAGRAM

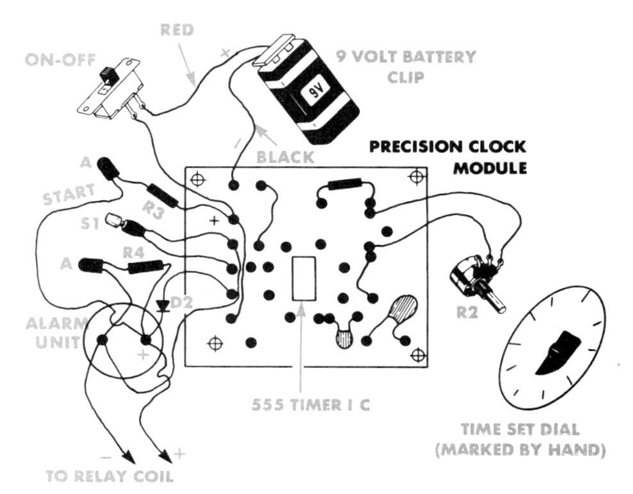

This timer is simple to build but accurate. It will sound an alarm or close a relay at the end of the time period. The relay can be used to turn appliances on or off. Look in the Construction Hints Section to see how to hook up relays. You must mark the dial by hand after you build the unit. The table shows how to find the time length you want.

SCHEMATIC DIAGRAM

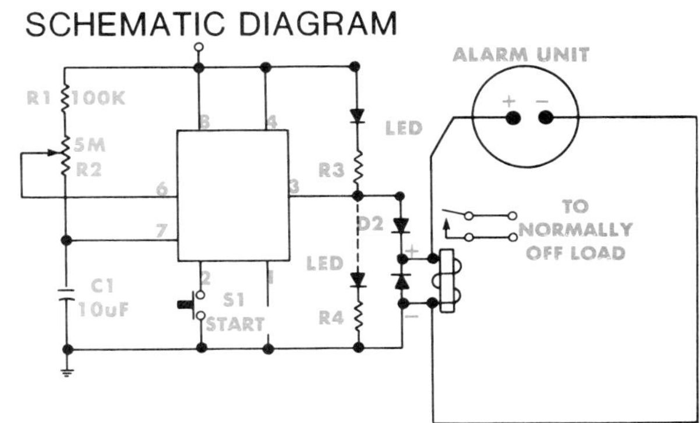

Parts List

1 Precision clock (timer) module
1 5 Megohm linear taper potentiometer (R2)
1 100,000 (100K) Ohm resistor (R1)
2 Light emitting diodes (LED)
2 330 Ohm resistors (R3 and R4)
1 Silicon diode
1 NORMALLY OPEN push button
1 Alarm unit and or 1 relay
1 SPST slide switch (ON-OFF)
1 9 volt battery and clip
1 Case (a plastic silverware drainer can be used for a case)

Finding the Value of C1 for the Time Period You Want

TIME PERIOD	VALUE OF C1 IN MICROFARADS
0 TO 10 SECONDS	.022
0 TO 30 SECONDS	.1
0 TO 1 MIN.	.22
0 TO 5 MIN.	1
0 TO 10 MIN.	2.2
0 TO 30 MIN.	6.2
0 TO 1 HOUR	10
0 TO 2 HOURS	22

NOTE: CAPACITORS FOR 5 MINUTES OR LONGER MUST BE TANTALUM TYPES

COUNTER MODULE

CIRCUIT BOARD LAYOUT

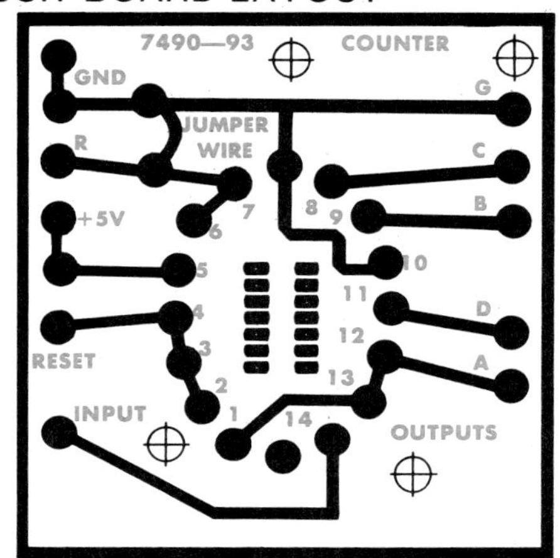

SCHEMATIC DIAGRAM

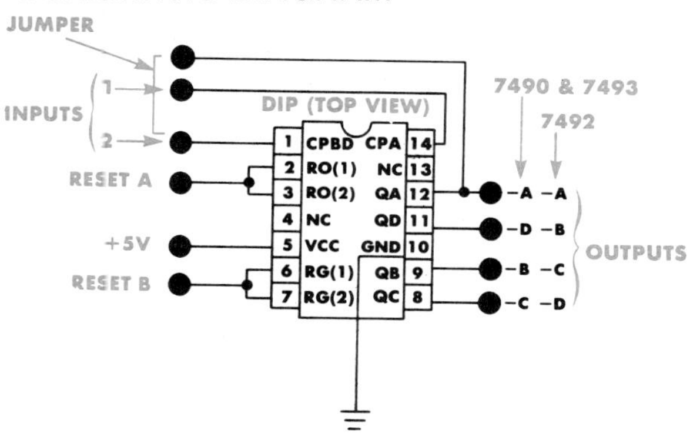

This circuit board is designed to use any one of three counting integrated circuits. One IC counter counts in binary 0 through 15, one counts from 0 through 9, and one counts from 0 through 11. The projects use either the binary or decimal (0-9) IC counter. All of the counters produce their count output in a binary code. That binary code is used in the lights out and super timer projects. The wheel of fortune project uses a DECODER integrated circuit to convert the binary outputs into a decimal form. These projects represent only a sample of ways to use counters.

Parts List

1 Circuit board
1 14 Pin DIP wire wrap socket for the counter IC
1 7490 Counter IC or 7493 counter IC (depending on which project you are building)

NOTES:

TYPE	COUNT	RESET A	RESET B
7490	0-9 (MOD 10)	1. +5 VOLTS RESETS COUNTER TO ZERO. 2. MUST BE GROUNDED FOR COUNTING	1. +5 VOLTS SETS COUNTER TO 9. 2. MUST BE GROUNDED FOR COUNTING
7492	0-11 (MOD 12)	NOT USED	1. +5 VOLTS RESETS TO ZERO. 2. MUST BE GROUNDED TO COUNT
7493	0-15 (MOD 16)	1. +5 VOLTS RESETS TO ZERO 2. GROUNDED FOR COUNTING	NOT USED

COUNT DOWN CLOCK FOR TIME LIMIT GAMES AND SPORTS

This project can be used to time games and events. The project is the same as the wheel of fortune circuit except that the run down clock is replaced by a precision clock module and a reset button is added. Use the table to find the right value for C1.

Instructions

Build the wheel display module and the decoder module as shown in the wheel of fortune game. Wire a counter module and a precision clock module as shown. There is a jumper wire between the counter reset terminal and +5 volts in the wheel of fortune circuit. That jumper must be removed for this project so the counter can be reset.

Finding the Value of C1

TIME FOR ONE COMPLETE WHEEL ROTATION. (ALL 10 LEDS)	VALUE OF C1
10 SECONDS	.68 MICROFARADS
20 SECONDS	1.5 MICROFARADS
30 SECONDS	2.7 MICROFARADS
1 MINUTE	4.7 MICROFARADS
5 MINUTES	22 MICROFARADS
10 MINUTES	47 MICROFARADS
20 MINUTES	100 MICROFARADS
30 MINUTES	120 MICROFARADS

CONNECTING THE MODULES

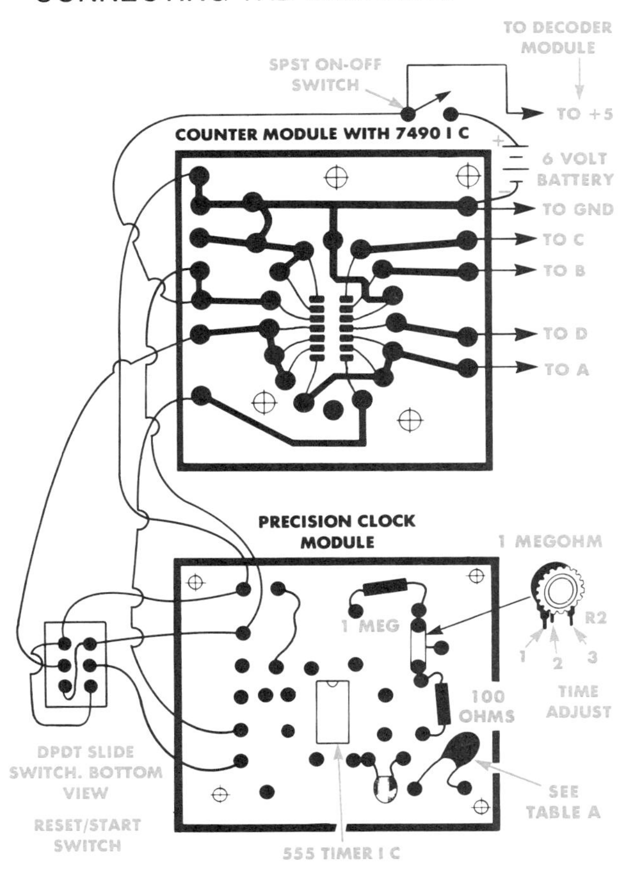

LIGHTS OUT GAME

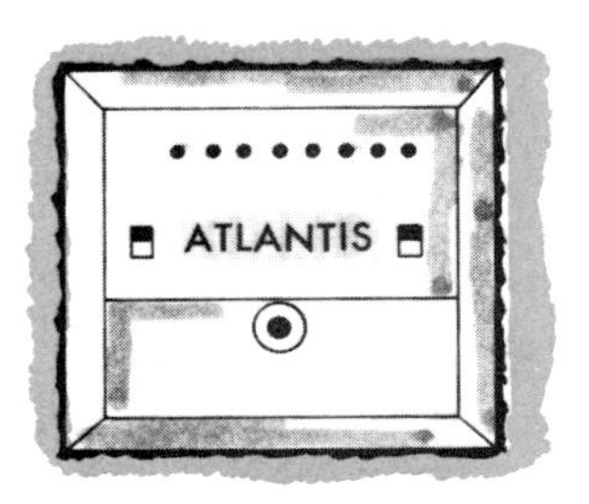

This game is believed to have been popular in the lost continent of Atlantis. You can believe that, too, if you believe in Atlantis. Anyway it is a fun game.

The game uses 2 counter modules and a special display module. If you play the game enough times it is possible to learn the code and win every time. To take care of that problem, you can use the "dressmaker's" plugs to scramble the code every so often. You will find drawings of how to make the plugs in the Construction Hints Section.

Parts List (Display Board)

1 Circuit Board
9 330 Ohm resistors
8 Light emitting diodes (LED)
1 SPDT or DPDT slide switch (Reset / Start)
1 NORMALLY OPEN push button
1 2.2 microfarad tantalum capacitor
1 SPST On-Off switch
4 Size C batteries
1 Battery holder
2 Counter modules with 7493 counter IC's
A case to put it in

CIRCUIT BOARD LAYOUT

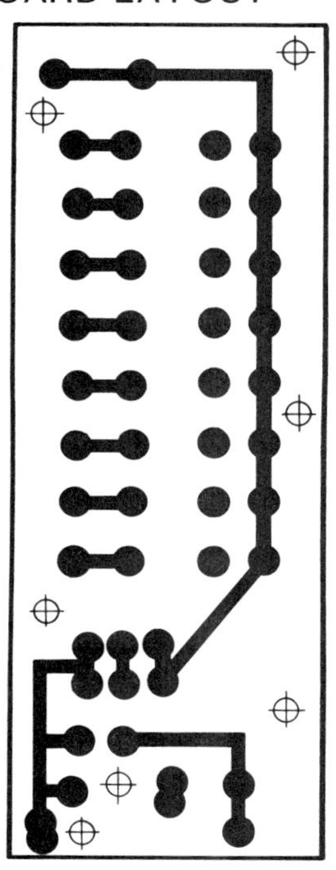

WIRING DIAGRAM

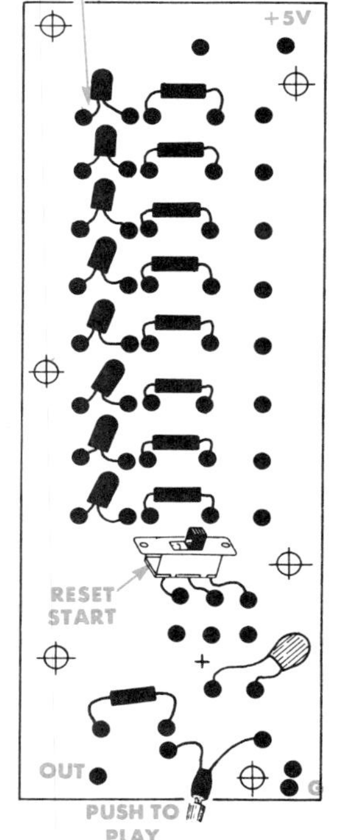

CONNECTING THE MODULES

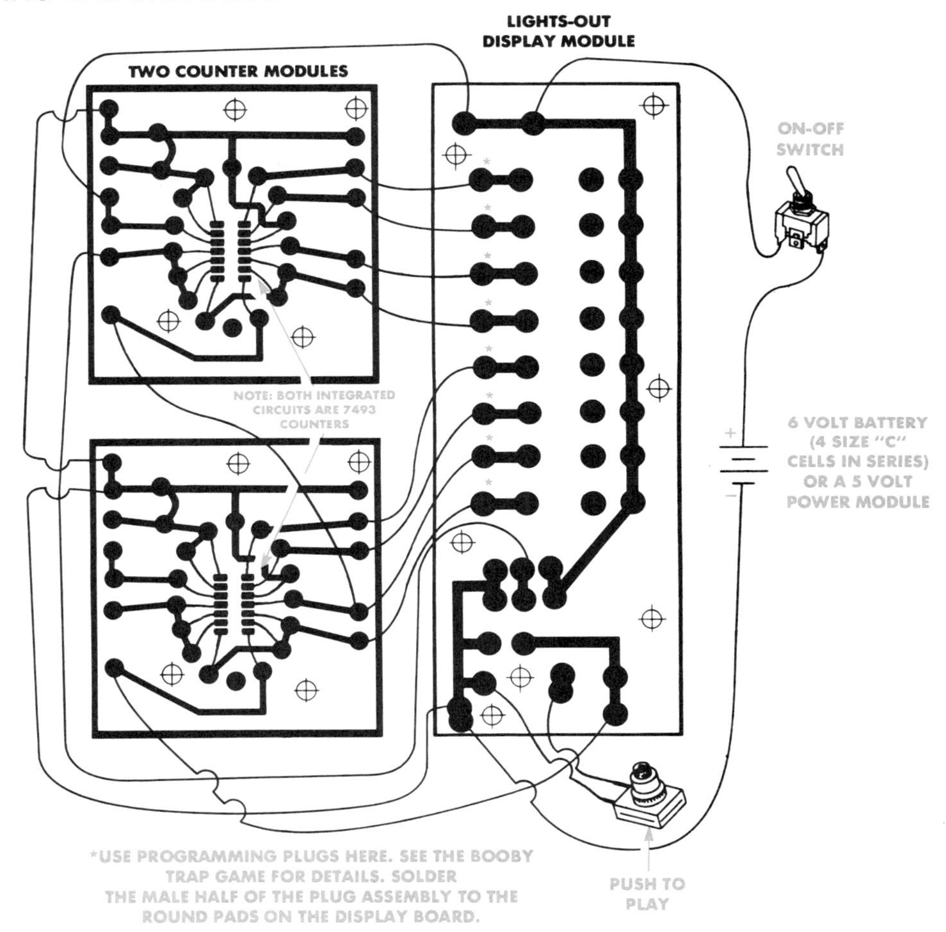

Playing the Game

1. Two players take turns
2. The game is reset to start. This turns on all of the LEDs.
3. Each player can push the button 1, 2, or 3 times. Pushing the button will turn out one or more lights. Pushing the button may also turn on one or more lights at the same time. For example you may turn one light off and two lights on. The player must say how many times he is going to push the button before he does it.

To Win the Game

Make your opponent turn out the last light. **The** player that turns out the last light has LOST the game.

WHEEL OF FORTUNE GAME

The game board uses a special wheel-like LED display board. The lights race around giving the illusion of a spinning wheel. A run down clock is used to make the "wheel" spin fast at first and gradually slow down to a stop. The "wheel" can be used with other board games or by itself.

Parts List (Display Board)
- 1 330 Ohm resistor
- 10 Light emitting diodes (LED)
- 1 Display circuit board

Parts List (Decoder Module)
- 1 Decoder circuit board
- 1 16 pin DIP wire wrap socket
- 1 74145 Decoder Integrated circuit

Parts List (Run Down Clock Module)
- 1 .47 Microfarad Capacitor (C1)
- 1 50 Microfarad Electrolytic Capacitor (C2)
- 1 100 Microfarad Capacitor (C3)
- 1 50 or 100K trimming potentiometer (R3)
- 1 NORMALLY OPEN push button (Push to spin)

Other Parts
- 1 6 Volt battery and battery holder
- 1 9 Volt battery and clip
- 1 SPDT or DPDT switch (ON-OFF)
- 1 Picture frame and wood top, or a box

CIRCUIT BOARD LAYOUT WHEEL DISPLAY MODULE

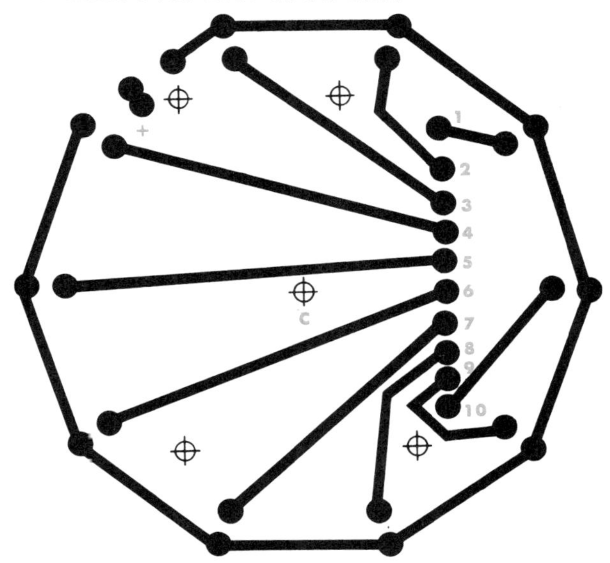

WIRING DIAGRAM WHEEL DISPLAY MODULE

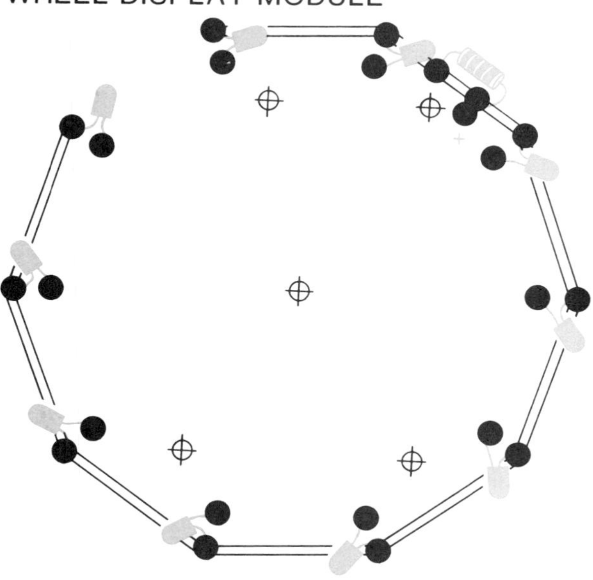

CIRCUIT BOARD LAYOUT
BINARY TO DECIMAL DECODER MODULE

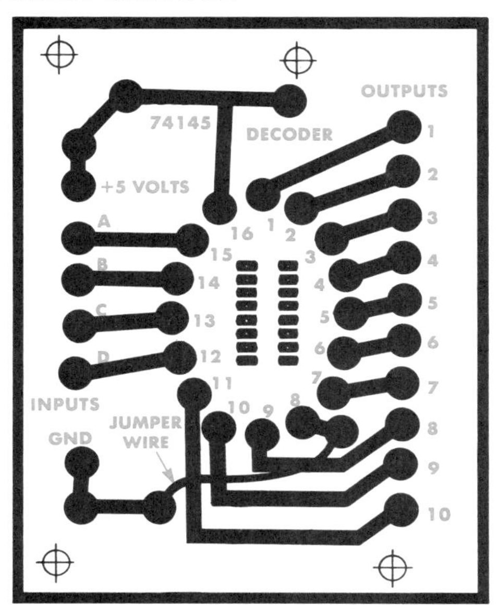

CONNECTING THE MODULES

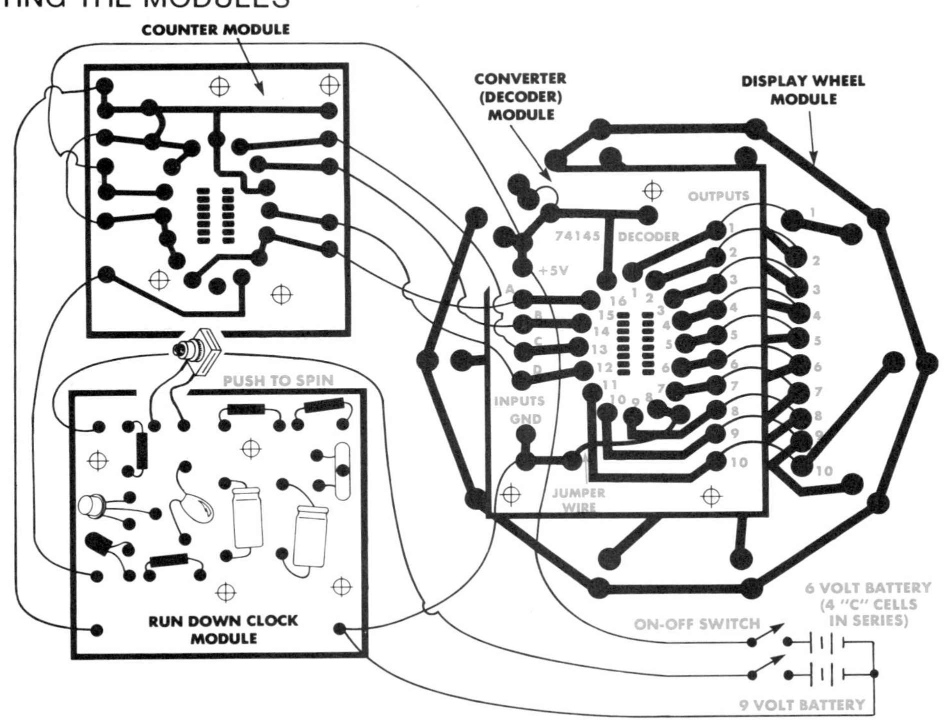

SUPER TIMER

This slick little timer has lots of applications. It can be set to the desired time range by selecting the proper capacitor for C1 in the precision clock module. The time is set with four slide switches. The time is then indicated by the LED indicators. Each 5 minutes (in the example) the lights change to show how much time is left before the alarm sounds. Values for C1 are shown in the table.

SCHEMATIC DIAGRAM

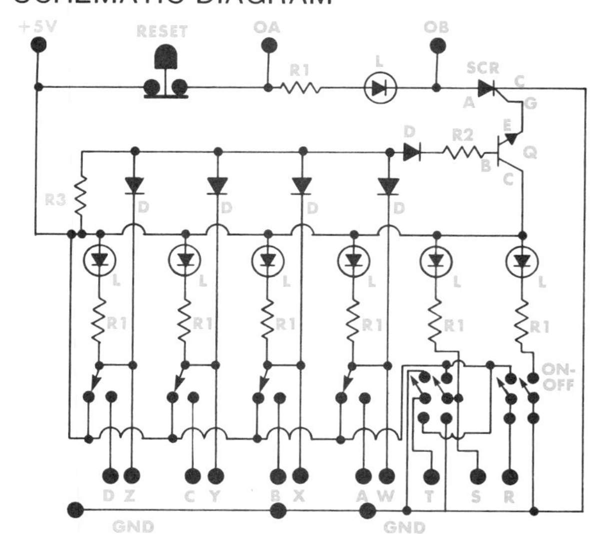

CIRCUIT BOARD LAYOUT
DISPLAY AND CONTROL BOARD

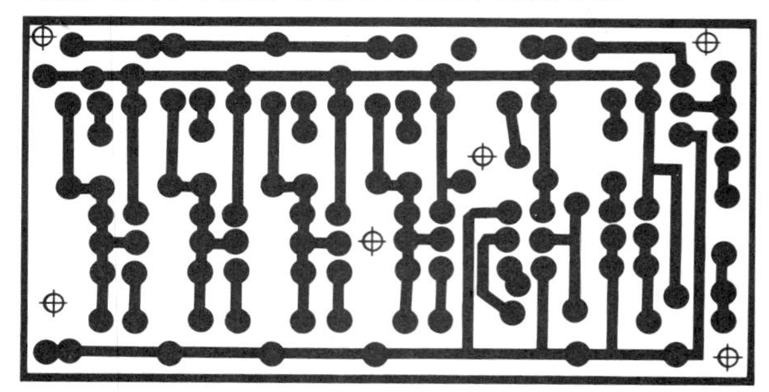

WIRING DIAGRAM
DISPLAY AND CONTROL BOARD

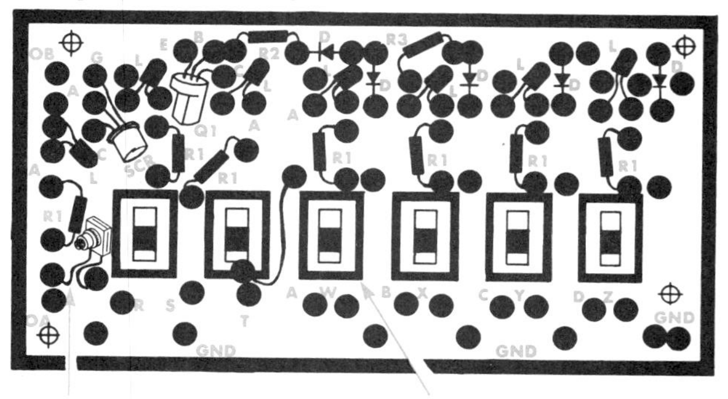

Parts List

1 Precision clock/timer module
1 Counter module with a 7493 integrated circuit

Parts List (Display Module)

1 Circuit board
6 DPDT slide switches (standard size, not miniature)
7 Light emitting diodes (LED)
7 150 Ohm resistors (R1)
5 Silicon Diodes (D)
1 NORMALLY CLOSED push button (reset alarm)
1 Solid state alarm sounder
1 Silicon control rectifier (SCR)
1 4700 Ohm (4.7K) resistor (R2)
1 100,000 Ohm (100K) resistor (R3)
4 Size C batteries and battery holder

CONNECTING THE MODULES

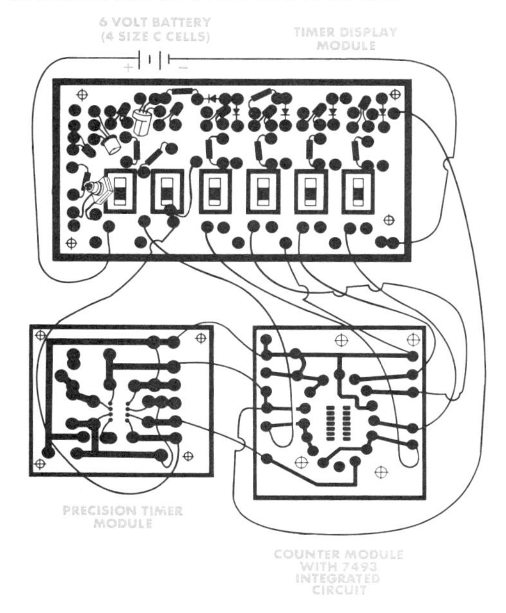

CONTROL PANEL

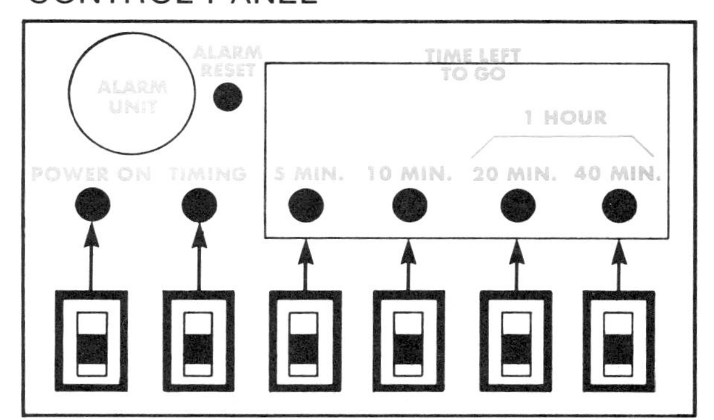

Finding the Value of C1

					VALUE OF C1 MICROFARADS
			1 MIN		
1 MIN: 15 SEC	5 SEC	10 SEC	20 SEC	40 SEC	.01
7 1/2 MIN	30 SEC	1 MIN	2 MIN	4 MIN	.062
15 MIN	1 MIN	2 MIN	4 MIN	8 MIN	.22
			1 HR		
1 HR: 15 MIN	5 MIN	10 MIN	20 MIN	40 MIN	1
7 1/2 HR	1/2 HR	1 HR	2 HR	4 HR	360
15 HR	1 HR	2 HR	4 HR	8 HR	1000

NOTE: RESISTOR VALUES FOR THE PRECISION CLOCK MODULE:
R1 = 3.3 MEGOHM
R2 = 2.5 MEGOHM POTENTIOMETER (TIME ADJUST)
R3 = 100 OHMS

MASTER PARTS LIST AND SUBSTITUTIONS

RESISTORS
All resistors are ½ watt carbon.

CAPACITORS
a. Values from .001 to .1 microfarad are ceramic disc types.
b. Values from .1 to 47 microfarad are tantalum types.
c Values above 47 microfarad are electrolytic or tantalum. Tantalum is preferred when available.

LIGHT EMITTING DIODES (LED)
Any visible LED will work in all of the projects.

SILICON DIODES
Any Silicon diode rated at 200 to 600 volts and from 1 to 3 amps will work in any of the projects.

SILICON CONTROL RECTIFIERS (SCR)
Any SCR rated from 50 to 600 volts and from 1 amp to 10 amps will work in any of the projects.

SPEAKERS
All speakers are 8 Ohms. Any size will do. A rating of ½ watt will do for anything except the main speakers for the stereo amplifier project. These should be rated for at least 5 watts (8 Ohms) and should be 6 inches or more in diameter.

TRANSFORMERS
All projects that use a transformer use a standard 6.3 volt 1 Amp transformer. It is often called a filament transformer. SUBSTITUTE: If the electronics type transformer is not available you can use a doorbell transformer from the hardware store.

RELAYS
Any 6 volt 250 to 600 Ohm relay will do.

ALARM SOUNDERS
Any sounder rated at 5 volts and 50 milliamperes (mA) or less will do. Try Radio Shack 273-049 or 273-060.

GENERAL PURPOSE (small signal) TRANSISTORS
Any NPN Silicon type that meets the following requirements can be used:
a. H_{FE} = 100 or more
b. V_{CE} = 25 volts to 100 volts
c. I_C = 50 milliamperes (mA) or more

Some Transistor Numbers
2N5210 (*RS 276-2013)
MPS2222A (RS 276-2009)
MPS 3704 (RS 276-2014)
MPS 3565 (RS 276-2031)
*RS = Radio Shack

POWER TRANSISTORS
Any Silicon NPN power transistor can be used if:
a. H_{FE} = 50 or more
b. V_{CE} = 25 Volts or more
c. I_C = 1 amp or more (up to about 15 amps)

Some Power Transistor Numbers
2N6569 (RS 276-2039)
2N3055 (RS 276-2041)
2N6576 (RS 276-2042)
D44H7

UNIJUNCTION TRANSISTORS
Almost any type will work. Try MU4891 or Radio Shack 276-2029

CADMIUM SULPHIDE PHOTOCELL
Try Radio Shack 276-116

PHOTO TRANSISTOR
FPT 100 or Radio Shack 276-130

GLOSSARY

Acid core solder: A kind of solder that should never be used in electrical work. It has a hollowed out center filled with acid.

Active circuit: A live circuit, one that has electricity running through it.

Ampere (amp): The unit of electrical current 6,250,000,000,000,000,000 electrons per second.

Amplifier: A device that controls a larger output power with a smaller input power.

Analog: Continuous—like a clock with hands as compared to a digital clock.

Anode: The positive terminal on a diode, SCR, or LED.

Armature (rotor): The rotating part of an electric motor or generator.

Binary: A number system using only zeros and ones for digits.

Capacitor: A small electrical energy storage device—something like a small rechargable battery.

Cathode: The negative terminal on a diode, SCR, or LED.

Clip lead: A length of flexible wire with an alligator clip at each end.

Clock: In digital slang, a pulse generator oscillator that controls the timing of the circuits.

Commutator: A sliding switch used to make connection to rotating coils in a motor or generator.

Component: Another name for an electronic part.

Conductor: A material that can carry electricity.

Coulomb: 6,250,000,000,000,000,000 electrons.

Current: Electrons in motion. The unit is the ampere.

Digital: In electronics, on-off devices as opposed to devices using a range of values.

Diode: A one-way electrical current "valve".

Domain: A small group of iron atoms that act like a miniature magnet.

Electromagnet: A magnet that requires an electrical current to make it work.

Ferric chloride: A chemical used to etch copper for circuit boards.

Field coil: The coil in a motor or generator that does not rotate.

Flip-flop: A digital memory device that can store a zero or a one.

Gate: A digital logic device.

Graphite: A form of carbon that is a good conductor and also self-lubricating.

Hertz (Hz): Cycles per second.

Hydroelectric: The use of flowing water to turn on an electrical generator.

Insulator: A material that can't carry electricity.

K ohms: 1000 ohms.

Lead: Another word for wire.

LED (light emitting diode): A special diode that produces light when current flows. Available in infrared, red, amber, yellow, and green.

Megohm: 1,000,000 ohms.

Microfarad: The most common unit for capacitance. 1 microfarad = 1/1,000,000 of a farad. Symbol: μf.

Normally closed push button: A button that opens (turns-off) a circuit when pushed.

Normally open push button: A push button that turns on (completes) a circuit when pushed.

Ohm: The unit of resistance.

Open circuit: A circuit that does not have a complete current path—no current flows.

Parallel circuit: A circuit with several independent current paths.

Picofarad: Unit of capacitance equal to 1/1,000,000,000,000 of a farad. Symbol: pf.

Power: The ability to do work. Power = volts × amps. The unit is the watt.

Programming: Telling a machine what it is to do and how.

Relay: A switch operated by an electromagnet.

Reset: To make a circuit go back to a starting condition.

Resistor: A component used to control the amount of current in a circuit.

Resonance: A special condition where two energy storage devices exchange energy at a definite rate with very little energy loss.

RMS: An average value for alternating current.

Schematic diagram: A diagram that shows how to connect electrical or electronic parts together. Each part has a special symbol.

SCR (Silicon Controlled Rectifier): A special diode that can be turned on by applying a signal to the gate element. An electronic switch.

Semiconductor material: A material that is not a good insulator nor a good conductor. Examples: carbon, silicon, germanium, gallium arsenide.

Series circuit: A circuit with only one current path.

Short circuit: Complete circuit with almost no resistance—causes too much current to flow.

Tolerance: The amount a part value can be off without the device being rejected.

Transformer: An alternationg current device used to step voltage up or down.

Transistor: An amplifying device made of semiconductor material.

Voltage: The electrical force that pushes current through a circuit.

Voltage gain: A figure of merit in amplifiers. Equal to output divided by input voltage.

Wire nut: A plastic nut for connecting wires together without solder.